SHAPING *the* UPPER CANADIAN FRONTIER

SHAPING *the* UPPER CANADIAN FRONTIER

Environment, Society, and Culture in the Trent Valley

Neil S. Forkey

UNIVERSITY OF CALGARY PRESS

University of Calgary Press
2500 University Drive NW
Calgary, Alberta
Canada T2N 1N4
www.uofcpress.com

National Library of Canada Cataloguing in Publication Data

Forkey, Neil S., 1964–
Shaping the Upper Canadian frontier: environment, society, and culture in the Trent Valley / Neil S. Forkey.

Includes bibliographical references and index.
ISBN: 1–55238–049–1

1. Trent River Valley (Ont.)—History—19th century. I. Title.
FC3095.T7F67 2003 971.3'6702 C2002-911377-6
F1059.T7F67 2003

We acknowledge the financial support of the Government of Canada through the Book Publishing Industry Development Program (BPIDIP) for our publishing activities.

Canada Council for the Arts
Conseil des Arts du Canada

Printed and bound in Canada by AGMV Marquis
♾ This book is printed on acid-free paper.

Cover design by Kristina Schuring
Interior typesetting by Jeremy Drought, *Last Impression Publishing Service*, Calgary, Alberta

for Roseline

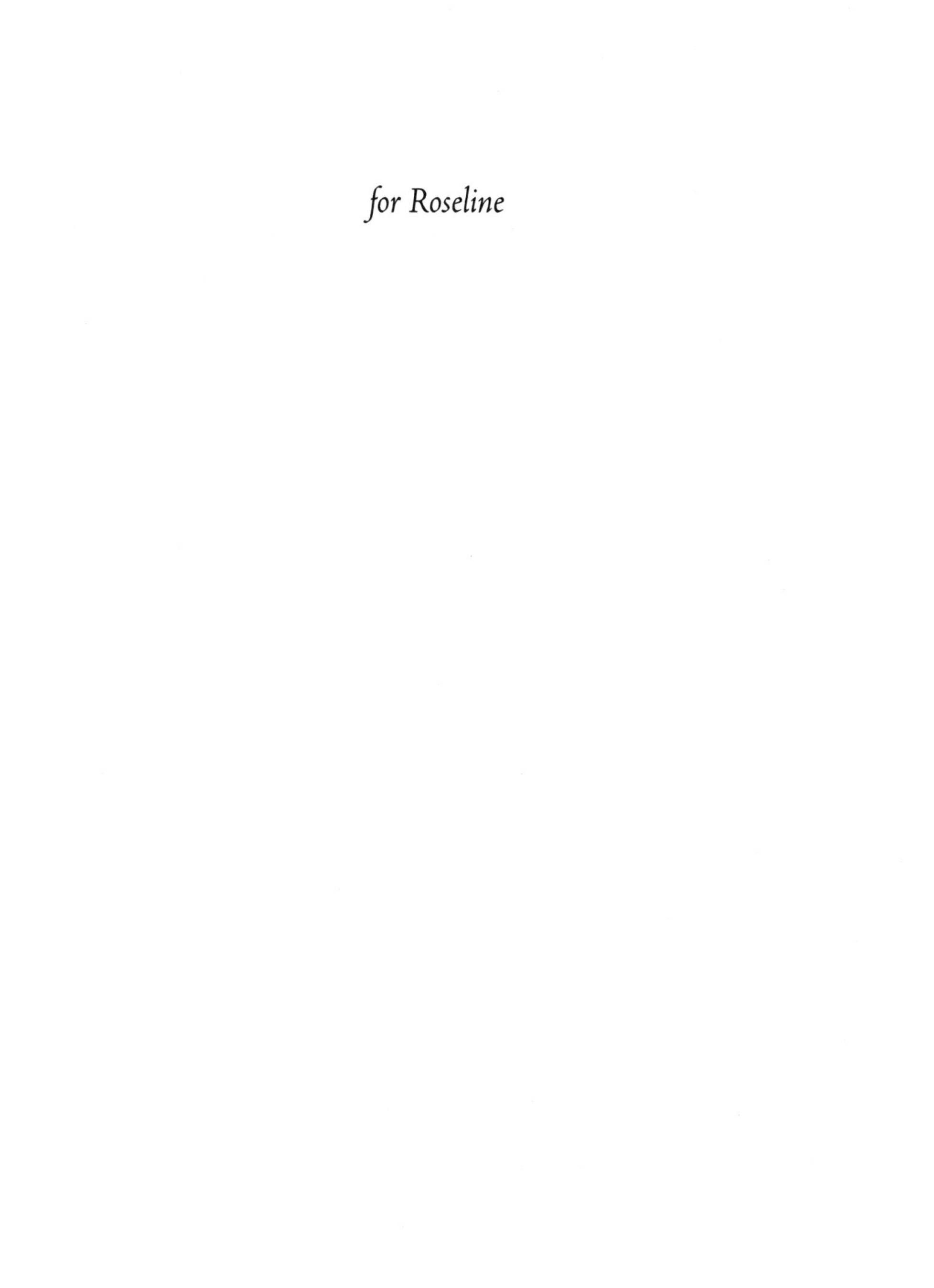

Contents

Part One: Foundations

1

2

3

Part Two: Contexts

4

5

6

7

List of Tables

List of Figures & Illustrations

Acknowledgments

Many people and institutions assisted my research and writing. A Donald S. Rickerd Fellowship, a Canadian Embassy (Washington, D.C.) Graduate Student Fellowship, a Joseph Leslie Engler Dissertation Fellowship in Canadian History, and funding from Queen's University helped to offset my expenses. This book has been published with the help of a grant from the Humanities and Social Sciences Federation of Canada, using funds provided by the Social Sciences and Humanities Research Council of Canada.

Leon Warmski, Jack Choules, Paul McIlroy, and Mary Ledwell at the Archives of Ontario; Patricia Kennedy and Trish Maracle of the National Archives of Canada; Bernadine Dodge and Professor Elwood Jones at Trent University, and George Henderson of the Queen's University Archives were all very helpful. Marilyn Croot (and Ela Rusak) produced the maps for this study. Joan Eadie indexed the book.

Professors Donald H. Akenson and Bryan D. Palmer were excellent doctoral thesis supervisors. Their patience and guidance was greatly appreciated. I also owe an intellectual debt to the following people, all of whom contributed to my understanding of Canada and the North American environment: Robert H. Babcock, Jacques Ferland, Richard W. Judd, C. Stewart Doty, William H. TeBrake, the late Edward Schriver, W. Fitzhugh Brundage, Geoffrey S. Smith, Brian S. Osborne, Suzanne Zeller, Robert W. Thacker, Joseph T. Jockel, and Alan M. Schwartz.

My family in Holyoke, Massachusetts and Montreal, Quebec—Dorothy and John Forkey, and Mariette and Jeremie Tremblay—have always been there to offer advice and support, while my wife, friend, and colleague, Roseline Tremblay, provided more encouragement that I can ever repay. For this and so many other reasons, I dedicate this work to her.

1

Introduction

In 1961 Hugh MacLennan heralded the seven "rivers that made a nation." The Mackenzie, the St. Lawrence, the Ottawa, the Red, the Saskatchewan, the Fraser, and the St. John are easily identifiable as great Canadian watercourses that, according to MacLennan, "link us with our past."[1] From a vantage point of more than forty years and from the purview of environmental history, one is struck by the author's bold assertion. His essays can be read as homage to natural systems that were forged during prehistory and that ultimately served as a stage upon which Canadian history was played out. From these vignettes one might conclude that MacLennan was an early proponent of what has been referred to as "bioregional" history: a narrative about a certain region or place written not merely from the perspective of human history but that also affords space within the dialectic for an analysis of the natural environment. Indeed, renewed interest in our environment and humankind's place within it has excited Canadian historians to delve deeper into the past.[2]

Taking a cue from MacLennan, I am concerned in this book with a much smaller Canadian river network found in southern Ontario, that of the Kawartha–Trent (once referred to as the Valley of the Trent, or here, the Trent Valley).[3] One will not find the Trent River or its sources and tributaries championed by MacLennan, yet the Trent River and the valley that bears its name does link Ontario (and by import, Canada) to its colonial past. The river nexus served as a conduit for peoples and ideas, which in turn catalyzed environmental change. In fact, it is my contention that the reciprocal relationship between people and environment that frames this study can serve as a model for understanding Canada's bioregional past.

The case of the Trent Valley bioregion during the nineteenth century suggests a microcosm for much wider human and environmental changes that were occurring throughout North America as the transplantation of European peoples sparked new relationships between humans and the new environments that they encountered.

In this respect we might view Upper Canada's development through an analytical lens similar to that recently suggested by Thomas R. Dunlap and in the collection of essays edited by Tom Griffiths and Libby Robin. These historians seek to unearth the relationship that English-speaking "settler societies" of the past two centuries had with their newfound environments.[4]

Bioregionalism might be understood in two ways. First, according to Carolyn Merchant, a holistic, postmodern definition might be "an idea that people and other living and nonliving things in a particular region, usually a watershed, are interdependent and that they should live as much as possible within the resources and ecological constraints of that place."[5] This might be an admirable yardstick for present societies, but of what use is this explanation when cast in historical terms? Dan Flores makes a cogent case for seeing American history as unfolding within particular bioregions, which then bear upon the national narrative. He proffers that the upshot of bioregionalism as a modern social movement "is not merely [a] focus on ecology and geography, but [an] emphasis on the close linkage between ecological locale and human culture, its implication [being] that in a variety of ways humans not only alter environments but also adapt to them."[6] Flores' suggestion for understanding a "sense of place" with the interplay of humans and their environment at the forefront is intriguing and affords the opportunity to emphasize on natural place and the human community.

Such an alternative will allow us to see the historical actors apparent in the Trent Valley, not the least of which is nature itself. It also affords us the opportunity to understand the motivations of the characters who appear in this study and grasp their conceptions of what the Trent Valley was, or what it could become. Put another way, we can appreciate the chasm between what William Cronon, in describing the birth and maturation of Chicago as a great American city, identifies as "first nature" and "second nature." The former refers to our understanding of natural systems, and the latter to the historic human impulse to redesign these systems, impose our will upon them, and thus recreate nature.[7]

Cronon's methodology is useful, but it must be adapted to fit the narrative of the Trent Valley. For Cronon it is an urban centre that commands the greatest attention; however, I seek to uncover such environmental change in a rural setting. In fact, this is a logical approach when considering a bioregion set in Upper Canada, which during the first half of the nineteenth century contained only three principal

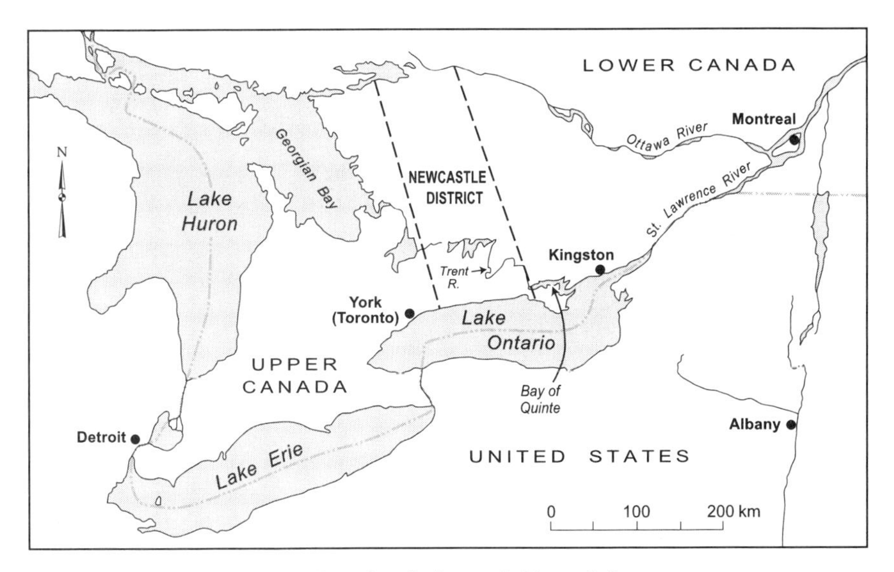

Figure 1: *Upper Canada showing the Newcastle district.*

urban areas—York (Toronto), Kingston, and Bytown (Ottawa). In the case of the Trent Valley, one must modify Cronon's model to account for the variables most accentuated in frontier settler life: clearing of forest, planting crops, and designing transportation routes that made the region habitable for agriculturalists. As the wilderness turned to rural landscape, and as a "home place" began to emerge, reconstitutions of first and second nature became evident, and it is this bioregional narrative that occupies my attention in these pages.

Ecological Locale:
The Trent Valley as a Living Place

Any bioregional approach to history must begin with the land and its features. The Trent Valley is a product of the last ice age. A mass of pre-Wisconsin glacial drift immediately north of Lake Ontario and running east to west formed the drainage divide south of Lake Scugog and Rice Lake. This divide further extends from the Niagara Escarpment in Caledon Township to the Trent Valley. The glaciers cut deep ridges, which dissect the watershed plain and give rise to several waterways

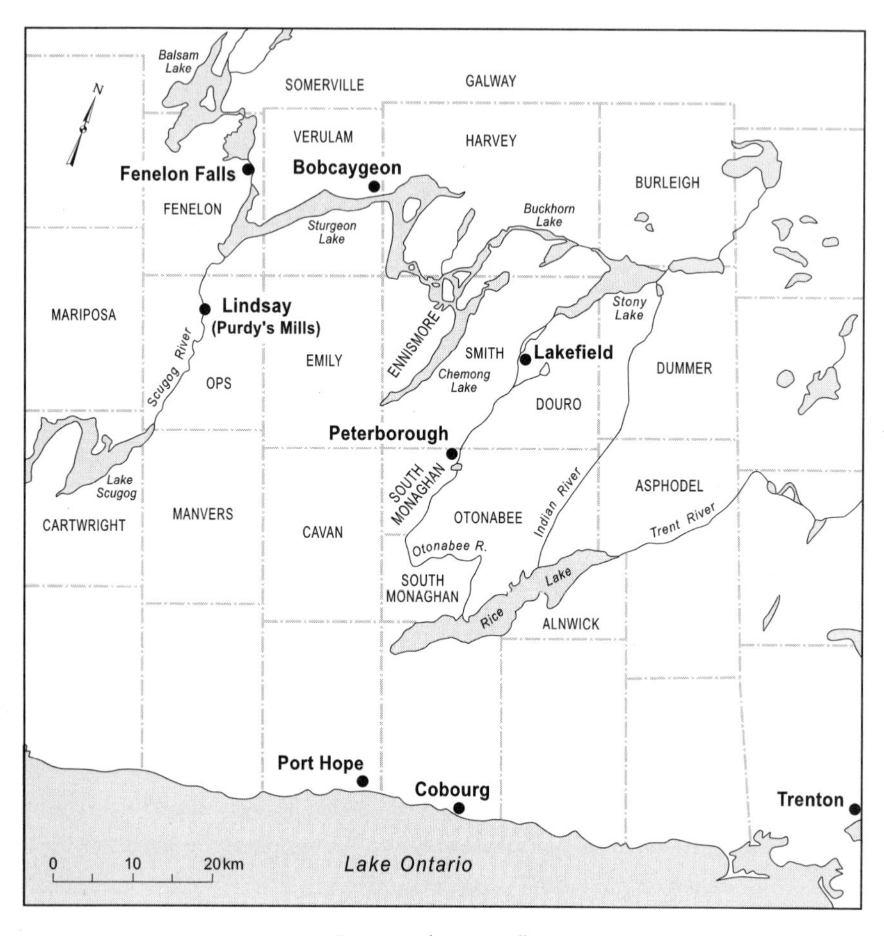

Figure 2: *The Trent Valley.*

containing wide swampy bottoms traversed by sluggish streams. Some watercourses, such as the Scugog River, act as drainage basins for the higher elevations.[8]

Geologically the Trent Valley is characterized by *drumlins*, Celtic for "little hills." Drumlins dot the southern Ontario landscape from north Hastings County to Lake Simcoe, with the Lake Scugog–Rice Lake axis comprising its centre. The Peterborough Drumlin Field sweeps across the lower Trent Valley from east of Rice Lake, skirting the southern shores of Chemong, Buckhorn, Pigeon, Sturgeon, and Balsam lakes (collectively known as the Kawarthas), to Simcoe County in the west. These rolling hills generally consist of a medium-textured soil and a mixture of clay, sand, pebbles, and boulders deposited centuries ago by the retreating glacier.[9]

Sandwiched between the drumlins and Lake Scugog are clay-based plains. The area's poor absorption foundation and its nearness to the drumlins combine to form imperfectly drained lands. Even within the past century, cropping has been difficult, the land being more suitable for pasturing livestock. During the first half of the nineteenth century, the area was characterized by swamp and mosquito infestation.[10]

The Dummer moraines are the third distinct surface type within the valley. Beginning at their namesake, Dummer Township, and bordering the Canadian Shield at the Kawarthas, the moraines denote a rough and stony land.[11]

From Galway and Verulam townships in the south to the Muskoka District, the area was distinguished during the nineteenth century by rich white pine forests and rugged land. Yet despite the drawbacks that hampered human mobility, the northward strip extending from Bobcaygeon north to Minden became a colonization route during the second half of the nineteenth century.

The Kawarthas and the two principal river systems, the Otonabee and the Trent, flow through the area. The Trent drainage area encompasses 4,790 square miles. The region's headwaters lie in the central part of Haliburton County and include nearly 130,000 acres of water surface. An additional 100,000 acres comprise nine other basins, principally those of the Gull, the Burnt, and the Mississauga rivers, which drain southwesterly into the Kawarthas. The overflow of these lakes is carried south into Rice Lake by the Otonabee River; the Trent River conveys Rice Lake's surplus to the Bay of Quinte.[12]

The valley's biogeography has been shaped by several factors. The area has approximately 135 annual frost-free days, and precipitation is approximately eighty percent rain and twenty percent snow. Such a situation gave rise to a prehistoric ecosystem dominated by maple, beech, basswood, and oak, with scatterings of pine on the more sandy soils and elm, ash, and white cedar where the soil was moist. As well as fur-bearing mammals, both large and small, varieties of birds are thought to have abounded in the thick forest. Fish such as lake trout and whitefish inhabited the more northerly and colder lakes, while perch and bass were most abundant in the warmer, southerly waters.[13]

Human Cultures:
The Settlement and Resettlement of Trent Valley

From prehistoric times to the nineteenth century, the Trent Valley was a main traffic artery between upper Lake Huron or Georgian Bay and the eastern end of Lake Ontario. As well, the region was the stage for rivalries over access to resources. The Kawartha–Trent nexus afforded relatively easy passage of several hunter-gatherer and more sedentary groups, all of which depended heavily upon the abundance of nature for sustenance.

The first known human group to inhabit the Trent Valley was the Late Archaic Indian culture more than thirty-seven hundred years ago, situated on the north shore of Rice Lake.[14] Following this group, two thousand years ago, the Hopewell culture occupied the southernmost portion of the valley, leaving to posterity the famous Serpent Mounds, a ridge two hundred feet from the north shore of Rice Lake and extending one hundred and eighty-five feet by twenty feet. The Hopewellians were agriculturalists, yet they enjoyed relationships with groups well to the east and south of the Trent Valley and left to archaeologists artifacts such as fresh water pearls from the Mississippi as evidence of their extensive trade network.[15]

Burial mounds approximately twenty miles north of Stony Lake at the Quakenbush site evince the Iroquoian presence of the fifteenth and sixteenth centuries, as do those at Le Vesconte on the Trent River, at Cameron's Point, and Rice Lake.[16] Indeed, the most unique remnants of the period of Aboriginal occupation are such mounds, as well as the petroglyphs—engravings in stone of animals and humans—at various locations in the valley. At Stony Lake, northeast of Peterborough, for example, the prehistoric Algonkians most likely inscribed in the limestone during the latter part of the Woodland period (c. 900 – 1400). The petroglyphs at Chemong Lake and Rice Lake are thought to have been left by the more recent Anishinabeg group, whose direct descendants are the modern Mississauga.[17]

Throughout this early period, Aboriginal peoples adapted to and altered the ecology to suit their needs. The Mississauga, for example, were particularly dependent upon the area's natural resources. In the Trent Valley they found plentiful flora and fauna upon which to subsist, as well as bountiful aquatic resources. They also grew corn and harvested copious amounts of wild rice from the shores of Rice Lake; even as late as 1817, ten thousand bushels of wild rice were collected annually.[18]

Aboriginal hunting was facilitated by burning large tracts of the area surrounding Rice Lake. The Mississauga selectively set fire to portions of the forest and underbrush so as to extend their hunting grounds. By stunting tree growth, the natives ensured a perpetual supply of grasses upon which deer and elk browsed. Thus, rather than embark on long forest excursions, the hunters lured the prey closer to camp. Forests were also burned to facilitate travel through often frequented corridors of the valley. These practices reshaped certain areas of forest for specific hunting-gathering ends, but eventually the process opened up vast areas such as the Rice Lake Plains. This area accommodated the nucleus of the eventual Anglo-Celtic community; Catharine Parr Traill, a member of this community observed in 1832 that "these plains were formerly hunting-grounds of the Indians, who, to prevent the growth of the timbers burned them year after year, this in process of time destroying the young trees, so as to prevent them again from accumulating to the extent they formerly did."[19]

In fact, the Mississauga name for Rice Lake, Pemadashdakota, translates as "the lake of the burning plains." So successful were the Mississauga that, over a period of about one hundred years, the once dense forest was transformed into the prairie and savanna-like Rice Lake Plains by the early nineteenth century.[20] Later, Anglo-Celtic pioneers would also use fire to clear land for permanent agriculture, although not always with the same precision: during the initial years of settlement, a fire swept uncontrollably through Peterborough shanties.[21]

The Mississauga relationship to and reliance upon the natural environment is discussed in greater detail in the next chapter. The remainder of the book is devoted to the shaping of the Trent Valley during the period of Anglo-Celtic occupation, which was made possible only after a substantial 1818 land surrender by the Mississauga.

In the Trent Valley, the Anglo-Celtic immigrants' initial land use focused upon forest clearance and agriculture. The enhancement of transportation and commercial farming was the goal of the Trent Canal project, begun in the early 1830s, which ironically took the early Aboriginal trade routes through the valley as a design inspiration. By the mid-1800s, lumbering had begun to occupy a prominent place in the local economy, with entrepreneurs ushering in an agro-forestry phase that continued into the twentieth century. All of these activities shaped the frontier of this Upper Canadian community and also informed and influenced the society that emerged after 1818. These aspects will be dealt with in a more systematic and detailed manner in subsequent chapters.

Nature and National Narrative in Canada

THE promise of a bioregional approach to Canadian history is that a natural and human sense of place might be better understood. It is not that the natural environment has been a neglected theme in Canadian historiography, in fact quite the contrary. Although sundry articles describing the introduction of plants and animals to Canada appeared during the first third of the twentieth century,[22] it was not until the Laurentian approach of the 1930s and 1940s that the environment emerged fully as a topic in Canadian historical scholarship. The study of environmental history, first in the United States and Europe and more recently in Canada, has again fixed attention upon this topic, but the more contemporary incarnation of the sub-discipline remains in its infancy.[23]

The environment and natural resources are prominent in the narratives of Canada's national historians. Harold A. Innis and Arthur R. M. Lower, for example, saw the colony and later the dominion's evolution firmly wedded to the international "staples" economy. Yet to consider the environment and its resources as mere inputs to production would be an obfuscation of nature's role in shaping human history. How then should Canadian environmental history treat the legacy of historians like Innis and Lower? Ostensibly, both were economic historians, yet natural resources were central to their work. As Carl Berger has noted, the perspectives of Innis and Lower cannot necessarily be considered in the same breath. While Innis clearly tended toward geographic determinism, Lower attempted to flesh out the human contexts within the milieu of the staples extraction.[24]

In his classic studies of the fur trade and the cod fishery, Innis advances a political economy of Canadian development that continues to influence social scientists.[25] Lower similarly pursued the history of the Canadian lumber industry through a staples approach. A later generation of forest historians continued to premise much of their research on Lower's model of forest sector trade as an expression of the political economy of continentalism.[26]

Despite the broad framework that the staples thesis *could* provide to environmental historians, the approach is not without its faults. The main problem may lie within the framework itself, which tends toward a rigid and mechanistic set of variables and assumes that international markets, rather than human agency, were at the root of Canada's creation. It also reinforces the idea that nature is a passive object and not an agent of historical change.

Nearly two decades ago, David McNally levelled a particularly cogent critique of the staples approach to Canadian political economy, one that also had meaning for social historians. He rejected the late-1970s trend among political economists such as R. T. Naylor and Mel Watkins to combine aspects of Innis and Karl Marx. McNally asserted that Innis' work embodies a crude materialism, which leads to a systematic neglect of the role of social relations of production in economic life. The result is a rigidly deterministic interpretation of economic history whose central feature is "commodity fetishism."[27]

McNally's assertion is important not only for historians of Canada's social relations, but for the environment as well. If environmental history is the study of human interaction with nature, then the approach should deal intimately with these two variables and demonstrate an understanding of both in their own contexts. By appealing to the staples model for direction, Canadian environmental history runs the risk of clouding such an analysis. The task at hand, it would appear, is either to cautiously expand on the broad staples framework most familiar to Canadian historians or to seek out other models such as have been developed in Europe or in the United States, or among international historians.

Innis and Lower do not pre-date the advent of environmental history through their emphasis upon natural resources in Canada's development. Their studies are not prototypical Canadian environmental history. They were concerned with exportable commodities, while environmental historians are interested in explaining the reciprocal and non-reciprocal aspects of the human–nature dynamic.

Occupying a generational space with Innis and Lower, and an eminent national historian in his own right, was Donald Creighton. He, too, moved Canada's environment to the front and centre of one of his most powerful narratives, *The Commercial Empire of the St. Lawrence, 1760–1850* (1937). Eloquently written and penetrating in its analysis, Creighton's masterpiece encapsulates the post-Conquest competition between agriculture and commerce in Lower and Upper Canada. In both analyses, Canada's environment is central to the ebb and flow of history. The Canadian Shield and the St. Lawrence River moulded the first European inhabitants' initial perceptions and expectations of their New World. Later, informed by technology and enhanced by international markets that accepted Canada's staples, humans overcame the obstacles to their movement and predominated on the northern half of the continent. Thus Creighton acknowledges the majesty of the Canadian environment but exalts the agency of humans in overcoming such impediments; however, throughout the narrative he waxes eloquent as he weaves the land and its attributes into the grand history of Canada.

Creighton's work embodies the Laurentian interpretation of Canadian history, the counterpart to Frederick Jackson Turner's "Frontier" thesis in the United States. Creighton writes, for instance, that the "Canadian Shield and the [St. Lawrence] river system which seamed and which encircled it, were overwhelmingly the most important physical features of the area. They were the bone and the blood-tide of the northern economy."[28] When introduced into the story, both the French and English found that "[t]hey had to live in and by the new world; and they were driven, by this double compulsion, to understand the possibilities of the new continent and to exploit its resources. They could escape neither the brutal dictates nor the irresistible seductions of North American geography; and in an undeveloped world the pressure of these prime phenomena was enormous and insistent. Each society, after long trial and error, had read the meaning of its own environment, accepted its ineluctable compulsions and prepared to monopolize its promises."[29] And monopolize them they did. "The dream of the commercial empire of the St. Lawrence," continues Creighton, "runs like an obsession through the whole of Canadian history; and men followed each other through life, planning and toiling to achieve it. The river was not only a great actuality: it was the central truth of a religion. Men lived by it, at once consoled and inspired by its promises, its whispered suggestions and its shouted commands; and it was a force in history, not merely because of its accomplishments, but because of its shining, ever-receding possibilities."[30]

Nearly simultaneous with the acceptance of the staples approach and the kindred Laurentian model, an ironic challenge to both emerged. W. L. Morton's 1946 qualification to both versions is subtle and revealing. "The implications of the Laurentian thesis," he writes, "are ... a metropolitan economy, a political imperialism of the metropolitan culture throughout the hinterlands."[31] In other words, the hinterland, a peripheral place, had as much to offer as a subject as did staple products or the mighty commercial elite of the St. Lawrence River. Morton sees Manitoba as the link between colonial and modern Canada, with the eastern influence traversing the Shield and the indigenous Metis shaping the region. Yet the land and resources are central to the story.[32]

Moreover, Morton recognizes that human agency was at work as Canada's environment was transformed. Of the Victorian era, he states that the "essential achievement of the Age was that of pioneering—the replacement of wilderness by civilization, forest by field, river by road, a cluster of wigwams by the pattern of towns."[33] Indeed, from the earliest European settlement "the chief and central task remained the creation of a habitat; the land had to be made livable. From this necessity arose the Canadian fixation on the Pioneer, a belief that the supreme act of history was to be the first, first in the conversion of wilderness to civilization."[34] Armed with science and technology, Canadians of the Victorian age crafted a "human abode" from this wilderness and, as they ventured West, sought "to make the land habitable, and responsive to human needs."[35]

J.M.S. Careless' "metropolitan" thesis, stemming from or in reaction to the Laurentian tradition, is certainly one of the more compelling models for understanding Canada's past.[36] Following from Creighton's emphasis on Montreal as the gateway to the Canadian West and commercial success, Careless suggests that other lesser, yet equally important metropolises, sprang up as human and capital migration progressed. For Careless, it was Toronto where the dynamic politics of business and agriculturalists fused to promote the opening of the Canadian West. Careless, in essence, acknowledges and informs what Morton mused about both the West and the technology and motivation needed to subdue this new environment. The drive westward began in Toronto but would impact Manitoba as well.

Why raise Careless' metropolitan thesis? Because it has influenced a prominent American environmental historian, William Cronon, whose *Nature's Metropolis: Chicago and the Great West* (1991) is premised closely upon Careless' model.[37] However, Cronon added another analytical layer, that of ecological change. The opening of the American West from Chicago—the metropolis that consumed resources and from whence capital inspired new economies in agriculture and cattle raising in the hinterland—enabled vast environmental change to sally forth across the Great Plains, as Chicago replaced far off New York as a regional commercial centre.

Finally, Careless is noteworthy for the way in which he influenced generations of Canadian history students, academic and non-academic, to think about Canada as a place. The text that educated thousands of high school and university students, *Canada: A Story of Challenge*, first published in 1953, sets in the mind of the reader a fundamental idea in its chapter one title: "Geography Sets the Stage."[38] The welcoming St. Lawrence River encouraged commercial activity and subsequent settlement, but the harsh and forbidding Atlantic Appalachians and Canadian Shield soon marked the bounds of habitation. In the drama of nation-building one must, as Careless reminds us, acknowledge the land, place, and setting. It would be an embellishment to suggest that Careless regarded the land as an "actor" in the way that a current environmental historian might. Nonetheless, the land is not only a presence, but an integral part of his narrative.

Why has Canadian environmental history developed so slowly? Possibly because until very recently there was no real need for a comparable American or European model to replace or augment the strong tradition of historical geography. Andrew Hill Clark's legacy has indeed been strong, although he is often overlooked as an early purveyor of Canadian environmental history. He laid a basis for historical geography beginning in the late 1940s with his study of New Zealand, which also speaks to concerns of the environmental historian.[39]

Early on, Clark realized the interplay of spatial and temporal variables in creating or altering natural environments. He asserts that, in tandem, history and geography offers the most dynamic analysis of how human cultures influence land use. Cultural factors are integral to explaining the settlement process.

Given the distinctiveness of the Maritimes, for example, Clark urges against an analysis that considers the three provinces as one region. Instead, each province needs to be seen as a separate entity with its own history and geography.[40]

Clark recognizes, for example, that in Acadia the biotic and cultural transfers from France played a key role in the development of agriculture and that subsequent English settlement brought a distinct mercantilist ideology to the colony after the fall of New France in 1760. The aim of his cultural analysis was to counter the rising tide of American neo-Turnerian scholarship of the 1960s, as well as to qualify the prevailing Laurentian thesis in Canada.[41]

While neo-Turnerians espoused an environmental determinism, Clark viewed Acadian settlement as a reflection of the lifestyle that was transferred from France. He found in Acadia low population densities, which accommodated a relative lack of fertile land, strong communal bonds, respect for indigenous peoples, and land reclamation schemes. Thus Clark revealed more continuity between Acadian and French culture and less of an evolution of a wholly separate North American outlook.

Laurentian school proponents, such as Creighton, maintain that settlement patterns rested on the proximity of commercial routes. Clark, however, notes that the main staples of eastern Canada, cod and furs, were only minor elements of the Acadian local economy. On the peninsula, trade was limited between Acadians and

Cape Breton Island's Scots and did not involve the creation of a commercial metropolis. Even in the large northeast region, Acadia was only marginally important to the New England economy during the colonial period.

It is important to emphasize these points because Clark was asking a question similar to one that an environmental historian might ask: why did Acadia look more like France than like England prior to 1760? Cultural transfer is the answer, and not environmental determinism. If the English, and not the French, had first inhabited Acadia, it might have looked like a region of England. Clark reminds us that, "we must never forget the degree to which Atlantic colonies of England or France in North America ... were, in fact, still cultural suburbs of Europe."[42]

Canada's land and resources have indeed influenced many of its historians over the past century. Yet the recent emergence of environmental history from other quarters, namely the United States, signals that the field in Canada remains underdeveloped and in search of definition. This is underscored by Ramsay Cook when he suggests that Canada was "an environment without a history."[43] Astute readers will no doubt recognize an implication in this last statement: *micro-environments*, rather than the *nation's* environment are useful points of departure for environmental historians in much the same way that "limited identities" suggested alternatives to the national historical narratives a generation ago.[44]

Cook's point is emphasized in a more contemporary sense by historical geographer Cole Harris' poignant assertion that Canada is less a cohesive geographic entity than it is an "island archipelago spread over 7,200 east-west kilometres."[45] Therefore, when one begins to write a history of place in Canada, one must bear in mind another fact presented to us by history: according to Harris, "[i]n our electronic age, it is worth remembering that local feeling in this country did not first develop with the province or, earlier, the colony, both political abstractions well removed from daily life, but with the settlement, the place where people lived and whose horizons they knew."[46]

Thus it may appear that bioregional approaches to Canada's past pave the way back toward the more specialized, regional studies that have characterized Canadian historical scholarship for the past thirty years. This is a safe and logical course, and one that allows the environmental historian to draw upon the wealth of information that these regional studies have produced as we construct the natural and human history of Canada. Even the pan-Canadian MacLennan recognized the utility of understanding regional spaces within the much larger national narrative. Moreover, studying the bioregional past presents us with information that we, at the start of a new century, might use in better understanding our relationship to the world that we inhabit. This, for so many reasons, is a worthwhile endeavour.

2

Changes in Mississauga Lands

Ecology and Economy, 1790s – 1830s

> Whereas, many heavy and grievous complaints have of late been made by the Mississaga [*sic*] Indians, of depradations committed by some of His Majesty's subjects and others upon their fisheries and burial places, and of other annoyances suffered by them uncivil treatment, in violation of the friendship existing between His Majesty and the Mississaga [*sic*] Indians, as well as in violation of decency and good order: Be it known, therefore, that if any complaint shall hereafter be made of injuries done to the fisheries and to the burial places of the said Indians, or either of them, and the persons can be ascertained who misbehaved himself or themselves in manner aforesaid, such person or persons shall be proceeded against with the utmost severity, and a proper example made of any herein offending.
>
> PETER RUSSELL
>
> *Proclamation to Protect the Fishing Places and the Burying Grounds of the Mississagas* [*sic*]
>
> (14 December 1797)[1]

THE period between the late eighteenth and early nineteenth centuries was significant in the environmental history of the Trent Valley. During this span, changes were effected in the relationship between people and nature. Here I borrow both a phrase and analytical framework from William Cronon's excellent study of New England during the "contact" period.[2] The Mississauga subsistence hinged upon a secure and open maintenance of the land and its resources. However, wider political changes on the North American continent imperilled their position within Upper Canada and invested colonial administrators with the power to oversee and protect Mississauga lands. As we saw in the introduction, the Trent

Valley had long been a place of human habitation, with various groups occupying the land and resources. The formation of Upper Canada as a place of permanent settlement ushered in a change in the Mississauga landscape, both physical and cultural, after they surrendered lands within the valley in 1818.

Resource conservation occupied attention in Upper Canada as early as 1797. The issue was important amidst the changing cultural landscape, and nowhere was this more evident than in the Mississauga apprehension toward English conceptions of property, and the consequences for their hunting and gathering subsistence. Although protection, specifically for Mississauga fishing grounds, had been proclaimed it proved ineffective. In fact, the Trent Valley Mississauga pressed for a greater defence of their fish and game several times between 1829 and 1835. Their concerns, however, were given only fleeting attention by the government during the province's fourth decade. Nevertheless, the persistence of the Mississauga in preserving their traditions is a testament to their reliance upon the valley's resources and juxtaposes their experience with that of the newcomer Anglo-Celtic groups.

Mississauga Lands

Access to the valley's resources was key to Mississauga subsistence. Indeed, they had shaped the valley through controlled burning, which changed the dense forest to savanna or plain that not only allowed for greater ease in travel and trade but also facilitated the hunting of various fur-bearing animals such as deer, wolf, bear, and beaver. As well, at the water's edge they fished and harvested maple sugar and wild rice. They tapped the trophic levels of the forests, fields, and lakes in adapting to and surviving within the Trent Valley ecosystems.

Maple sugar and wild rice added greatly to the diet of the Mississauga. Anthropologist A. F. Chamberlain notes that families throughout the area drew from a sugar bush and that each spring Mississauga women gathered and prepared maple sugar.[3] As well, Emma Jeffers Graham fondly remembers her youth at the Hiawatha reserve bordering Rice Lake, where her father was the chief missionary. The lake was so named for the wild rice that grew there, attracting great flocks of wildfowl, which were then hunted. Each autumn the natives paddled to the beds and harvested the rice. "The Indian," she recalls, "possessed himself of two shingles, and with these scraped the rice from the stalks into the boat. When the rice was all

Figure 3: Gerald Sinclair Hayward, *Rice Lake, Ontario*, 1918, National Archives of Canada, C-129703.

gathered in, it was poured into large wooden pans where the Indian shelled it by trampling it (with new moccasins on their feet). The peculiar flavour of the wild rice was remarked by all of us, and we soon came to regard the white rice as tasteless."[4] Jeffers also remembers the "great stir in the village," engendered by the start of the fall hunting season; the "spoils of victory" would sustain the Mississauga during the winter months.[5] Reginald Drayton also recalls the September rice harvests. He lived along Rice Lake near Hiawatha and recounts that "the wild rice by reason of which the lake is named ... grows as well ... wherever there is a good black soil and the water is not too deep. This rice which is excellent to eat & can be cooked in various ways used to belong exclusively to the Indians and they found a ready sale for it with the white people."[6]

The wildfowl of which Jeffers writes also attracted the attention of another commentator, Catharine Parr Traill, who arrived in the region in 1832. She observed that following the spring ice melt on Rice Lake, Mississauga families would pitch tents and hunt the ducks that had arrived to eat the wild, green rice. Hunters would camouflage their canoes with green boughs, approach the ducks, and shoot them. Not surprisingly, sportsmen from Traill's own cultural group mimicked the technique with great success.[7]

The Kawartha–Trent water nexus also provided a variety of aquatic species that served as year-round mainstays of the Mississauga diet. Throughout the watershed, the natives captured white fish, salmon, bass, trout, masquinonge, eel, and other fish, using several modes of capture including spear, seine, and gill-net. The adoption of such native technology enabled Anglo-Celtic newcomers to catch fish themselves and thus avoid purchasing such bounty from the natives. One of these new arrivals, Thomas Need, accompanied Mississauga fishers on their excursions. "This evening I accompanied an Indian on a fishing excursion," he wrote in April 1834. "[H]is tall figure, stationed at the head of the canoe, with the full glare of the torch thrown upon it, would have been a fine study for an artist. At the first lunge of his spear, a maskalongy [*sic*] or pike of 30 lbs. weight was secured, to which several others in turn succeeded, until at length our canoe was quite full."[8] Traill reported similarly from Rice Lake, where the natives engaged in night fishing with spears and jack-light in the summer months. Later in the year, when the lake froze, she likewise noticed numerous ice-fishers as they landed masquinonge and other species.[9]

Thus, in a variety of ways the Trent Valley ecosystem supported the Mississauga, the "human culture" in tenancy of the environment as Upper Canada was undergoing its initial formation. New challenges lay ahead, however, as the province underwent an infusion of new peoples, initially refugees from the new American Republic to the south and groups from the British Isles. What occurred province-wide was the introduction of new land-use regimes and conceptions of second nature which markedly altered the world just described. However, far from being passive characters in this drama, the Mississauga resisted the breaching of their traditional activities and both adapted to and persisted in maintaining their connection with the environment.

Resistance

FOLLOWING the American Revolution and the subsequent movement of Loyalists to British North America, as well as the increased levels of immigration from the British Isles early in the nineteenth century, Upper Canada, including the Trent Valley, became home to non-aboriginal peoples. One estimate sets the change in concrete terms: from 1815 to 1824 the non-Aboriginal population doubled from 75,000 to 150,000. In order to provide land for these newcomers, Upper Canadian administrators concluded six major land-cession agreements that brought almost three million more hectares under their control.[10] The agreement that specifically affected the Trent Valley was struck in November 1818. In return, 740 pounds were to be paid annually to the Chippewa Nation (of which the Mississauga were a part) and equally divided to each person.[11]

This change to the cultural landscape brought conflict. Disputes soon arose over the full meaning and implication of the land surrenders. Like many hunter-gatherers, the Mississauga did not conceive of nature's bounty in terms of private property, although they did draw a distinction between their territory and that of other Aboriginal and Euro-American groups. Their seasonal mobility dictated a "usufruct" approach to land. In essence, usufruct rights were predicated on the assumption that no one family or tribe owned the lands on which they hunted, fished, or farmed; such grounds might be accessed during a particular season, but it was understood that permanent settlement or over-exploitation of the resource was prohibited. Furthermore, in order to promote the natural regeneration of a particular hunting/fishing ground or of farmed land, native peoples would leave such sites untouched or fallow for several years. Such a place-based system best accommodated the Mississauga. As well, cumbersome material possessions, which could include stockpiles of food and furs, confounded their seasonal mobility and were thus deemed impractical.

The Mississauga subsistence regime may well have contributed to the European view that the natives did not "own" the land because they were not physically inhabiting a specific locale but rather transitionally encamping at several sites throughout the year. William Cronon suggests that in New England (and it may be inferred in Upper Canada also), such mobile Aboriginal lifestyles enabled Europeans to rationalize or justify the appropriation, or bounding, of native lands for their own permanent settlement.[12]

An example from the Village of Marmora is illustrative of this scenario. In the early 1820s the eastern branch of the Trent River system was in large part a wilderness. Charles Hayes ventured into this backcountry and constructed an iron works. While blazing a trail thirteen to fifteen miles from the Trent River to Marmora and erecting the mill, Hayes and his workers encountered native resistance. The Mississauga were angered by the workers' incursions onto their time-honoured hunting grounds. In March 1821 Hayes informed the lieutenant-governor's secretary of a worsening situation, which he feared might end in violence. Hayes discerned that the Mississauga annual payment for their ceded lands had yet to be paid, and they were especially perturbed in light of the recent encroachment upon the valuable hunting grounds around a beaver dam. The situation so concerned Hayes that he sought the "power" to counter a native attack and requested "30 stands of arms" with which to equip his men.[13]

The situation in Marmora had only partially subsided by the spring. Hayes still awaited the "presents" to mollify the natives, but had received instructions for repelling an attack: rather than arm his workers, he was advised to call upon the local militia for assistance. Hayes remained cautious, however, reporting, "I am happy to tell you [that] matters are wearing a more tranquil appearance and if presents were made I think the Indians would be quiet."[14]

The predicament may well have defused itself as the Mississauga simply migrated to other locales when the spring run of salmon and other anadromous fish commenced on waterways connected to Lake Ontario. Nonetheless, Hayes' concerns provide insight into an early and central issue of contention: permanent settlement was marginalizing the Mississauga on lands central to their traditional hunting-gathering regime.

By 1829, the Mississauga more adamantly asserted their concern for protection of their resources. Unlike the situation seven years earlier in Marmora, in January 1829 the Rice Lake area chiefs petitioned the provincial House of Assembly for protection of hunting grounds. Strangely and without explanation, however, the petition was withdrawn by legislators the next day.[15] Eight months later, in September 1829, Peter Jones, a Mississauga and a Methodist missionary, wrote to Commissioner of Crown Lands Peter Robinson. Evidently the encroachment problem had not subsided. At the request of the Rice Lake and Mud Lake chiefs and hunters, Jones explained that white hunters were encroaching on Mississauga land and taking game. Through Jones, the chiefs and hunters unsuccessfully petitioned Lieutenant-Governor Sir John Colborne in the hope that their lands would be better protected.[16] The problem persisted, and in January 1830 Chief Paudash and six other unnamed Aboriginals of Rice Lake petitioned the Provincial House of Assembly for protection of their hunting grounds from "wanton aggression." The petition was referred to a select committee, and two months later that committee was prepared to submit a draft bill, "whenever the House would be pleased to receive [it]."[17] The Assembly, however, did not proceed on the matter, which necessitated the filing of yet another petition. The next year, 1831, the natives again petitioned, only to have their concerns not acted upon.[18]

In 1831, Thomas Carr of Peterborough suggested ways to safeguard Aboriginal hunting grounds. The grounds around Mud Lake extended approximately seventy miles from the rear of the village. In the previous season, Mississauga grounds were "invaded by a band of [thirteen] lawless white hunters from the States and this neighbourhood." Carr recounted that in stark contrast to the conservation-minded natives, these Americans "in their wanton depradations [*sic*], spared neither age nor sex among the beavers; leaving, in short, none 'to keep seed alive.'" Moreover, the furs taken were valued at $600, which should have rightfully been given to the natives.[19] Mississauga concern seems to have been validated by the 1833 lamentation of Thomas Need regarding the scarcity of beaver in the area.[20]

Similar problems arose when the Mississauga specifically called for an act imposing a close time on Rice Lake masquinonge fishing. Yet this 1833 petition, too, fell on deaf ears in the Assembly.[21] An 1835 petition further delineated the Aboriginals' goals. In January 1835 a petition of Paudash and twenty others from Rice and Mud Lakes was tabled in the Assembly respecting "the danger that exists of the game and fish becoming extinct in the grounds and waters reserved for [the natives] in the original treaty."[22] American hunters and fishers continued to trespass upon Mississauga land with impunity. This was occurring despite the fact that neighbouring tribes had agreed to leave the lakes undisturbed for the next three years in the spirit of conservation. The Rice and Mud Lakes people now wanted a law passed by the Assembly that would check the actions of Canadians and Americans; they went so far as to recommend close times between the first day of April and June toward that end.[23] Despite the care taken by the natives to articulate the problem and to propose a solution, the petition was never acted upon, and the issue became a dead-letter.[24]

The Mississauga concern notwithstanding, one settler pointed out that the Aboriginals did not respect already existing game laws. In July 1835 Thomas Need noted that "according to the Provincial Game Laws, deer hunting begins in July and ends in February but this law, like many others in a growing Colony, is only of use inasmuch as it serves as a moral restraint upon the gentlemen: the mere backwoodsman regards it lightly enough, and the Indian not at all."[25]

Despite the lack of a favourable outcome, the petitions brought by the Mississauga evince a movement to protect the resources that were so dear to their subsistence. Peggy Blair's recent scholarship suggests that the Trent Valley Mississauga's assertion of their rights was consistent with the trend throughout Upper Canada, as other Aboriginal groups sought to resist the competition for resources on the newly surrendered lands.[26] This, as well, fits with the wider continental context demonstrated by Elinor G. K. Melville in pre-modern Mexico, where within one generation Aboriginals began to petition against Spanish pastoralists following the destruction of their crops.[27] Indeed, the alacrity with which the Mississauga sought to counter the threats to their way of life speaks to their reciprocal relationship to the land.

While on a missionary fund-raising tour of England, Jones further wrote about the Mississauga concerns. "As our people are growing wiser," he asserted, "they are much pleased that our great father [Lieutenant-Governor Colborne] is taking a new way with us, and giving us useful things as presents, and that the firewaters [liquor] is no more given us."[28] More to the heart of the matter, however, Jones expressed the problem of accessibility to the land. He and other Mississauga saw that the "country is getting full of the white people, and that the hunting will soon be destroyed. We wish our great father to save a sufficient quantity of land for ourselves and our children to live on and cultivate. It is our desire that whatever lands may be marked out for us, to keep the right and title ourselves, and not be permitted to sell them, not to let any white man live on them unless he is recommended by our own council, and gets a license from our father the governor. ...[W]e wish to feel that we stand on our own lands that our fathers left to us."[29]

Yet despite such resistance and desire for land tenure, adaptation seemed to be the only safe means by which the Mississauga could remain in the Trent Valley. This, however, was also a difficult task.

Adaptation and Persistence

During the mid-1820s Upper Canadian administrators began a program to marginalize the province's Aboriginals on allocated reserve lands.[30] A concerted effort was made to replace hunting-gathering activities, which were obviously clashing with new European settlements with sedentary agricultural pursuits. Moreover, in an effort to further "civilize" the natives, reserve superintendents, or in the case of the Rice Lake area Mississauga, Wesleyan Methodist missionaries, attempted to Christianize the Aboriginals. The Trent Valley Mississauga began their reserve inhabitation in 1828.

By 1843 the valley's Mississauga were living on four reserves: Alnwick, Rice Lake, Mud [Chemong] Lake, and Balsam Lake. They numbered 233 and, according to their superintendent, the Reverend William Case, they had taken satisfactorily to industry and agriculture. Each family had cleared at least half of their allotted twenty-five acres, a total of between 360 and 400 acres. They also kept livestock. However, Case lamented their "[fondness for] roving, by which the best season is lost for farming."[31]

Again Thomas Carr commented on the Mississauga, this time reporting on their transition to more permanent habitation. He also intimated that they had become more accepting of permanent white settlement and were moving toward a European cash economy. Carr and a friend visited the Mississauga village at Mud Lake, northwest of Peterborough. He observed that Mud Lake abounded with masquinonge, eel, bass, and waterfowl, and that the Mississauga there were engaged in rudimentary commercial fishery with the surrounding settlers.[32] Carr had learned that "the fish, from the influx of settlers are now in great demand." The clearings of Smith Township on the adjoining east side of Mud Lake provided a visible market. Carr asked an old native man about the fish prices. The man responded, "One dollar one. What do you say? One dollar one fish; fish scarce; weather cold." The Mississauga had evidently understood settlers' reluctance to venture into ice fishing and had inflated the price of the catch.[33]

Catharine Parr Traill was similarly impressed by the Mississauga's acumen. She remarked that locally grown cranberries, sold in town by the Aboriginals, formed the centrepiece of many a village table.[34] On one of her frequent trips to local encampments, Traill admired the women's decorative crafting of clothing from deer skin. When Traill gestured that she would like a few porcupine quill tools for herself, an elder woman gave her "a few of different colour that she was working a pair of moccasins with, but signified that she wanted '*bead* to work moccasin,' by which I understood I was to give some in exchange for the quills." "Indians," Traill observed, "never give since they have learned to trade with white men."[35]

Carr reflected on the natives' situation both within the Trent environs and throughout North America. He asserted that the efforts of the Mississauga's superintendent, Reverend Scott, an agent for a benevolent society in England, to civilize the natives was a worthwhile endeavour. Carr believed that "[a]mong the various progressive or contemplated improvements, which give animation to this fine [Newcastle] District, the civilization and settlement of the Indian tribes ... cannot be considered as one of the least humane or interesting.... A friend to humanity may therefore rejoice that the reign of systematic violence and oppression [against the Indian] has at length yielded to the exercise of the benevolence and generosity."[36] It might be surmised from Carr's comment that the reserve system had acted to protect the natives. In the past they had confounded European ideas of progress and been destroyed, either indirectly by disease or intentionally by force. By removing them from the busy path of North American progress, circumscribing their activities within reserves and, through religion and education, assimilating them into a European culture, they might be spared, or so reasoned some in the newcomer group.

However, the Mississauga only partially embraced the boundaries of the reserve system and agricultural pursuits, despite the promotion of their white superintendents and neighbours. In fact, a traveller remarked that seasonal foraging continued to occupy the natives' attention in spite of the reserve boundaries, and he urged the "tourist to turn aside from the beaten track to visit the Indians in the bush" so as to witness the Mississauga during autumn when "their hunting spirit breaks forth ... casting off the trammels of civilisation."[37]

Yet the historical record reveals much more than backward glances and expressions of lament for the passing of the Mississauga dominion over the Trent Valley. The evidence suggests that change certainly did occur within the Mississauga lands, but that the natives also displayed adaptive qualities by which they became participants in shaping the new era of land use.

Toward Resettlement

NEWCOMERS to the Trent Valley received a "ready-for-settlement" landscape from the Mississauga. The natives' subsistence mode had not permanently altered the area ecology, leaving abundant fish and game. Mississauga usufruct practices were intended to maintain a long-term balance with nature, ensuring a continual yearly yield of fish and wildlife. As well, the use of Aboriginal fishing technology helped to provide essential nutrients to the newcomers' diet during the initial settlement phase.

Mississauga habitation, which often included burned over and cleared forest, unwittingly facilitated later permanent European settlement. Once treaties were concluded and township surveys completed, Upper Canadian administrators began to earnestly populate the Trent Valley, thus shifting the economy from hunting and gathering to agriculture. Indeed, the advent of a "settler society" would prompt new relationships between humans and the land.

The Anglo-Celtic migrants would pursue more materialist goals and in the process try to replicate the landscape that they had left behind in the Old World. Clearly, another period of cultural resettlement was about to be visited upon the ecological locales of the Trent Valley. Moreover, as Anglo-Celtic groups carved a home in the forest frontier, distinct cleavages between first and second nature began to emerge, which stood in stark contrast to the conception of the Mississauga.

3

Creating New Home Places: *Anglo-Celtic Migrants,* 1820s – 1850s

Every little dwelling you see has its lot of land, and consequently, its flock of sheep. ... Many of these very [Upper Canadian] farms you now see [in 1832] in so thriving a condition were wild land thirty years ago, nothing but Indian hunting grounds. The industry of men, and many of them poor men, that had not a rood of land of their own in their own country, has effected this change.

An unidentified woman quoted in
CATHARINE PARR TRAILL
The Backwoods of Canada: Being Letters from the Wife of an Emigrant Officer, Illustrative of the Domestic Economy of British America (1836)[1]

A bioregional approach to history offers the opportunity to view the adaptation of human culture to an environment. The case of the Trent Valley during the nineteenth century allows us to gauge the changes to Mississauga lands, and a similarly useful exercise when probing the history of Upper Canada is to analyze the adaptive qualities of European immigrant groups. In this case, it was Irish settlers who undertook to make the Trent Valley their new home, or in bioregional parlance their "home place." These Irish migrants joined with families from England in shaping the ecology and creating a frontier community.

Anglo-Celtic peoples sank roots into the Trent Valley soils in their search for new homes and economic opportunities. In this process, they acquired local knowledge that enabled them to adapt and live in the valley. The migrants gained adaptation skills both from their attachment to the Irish landscape and from their observation and practice in the New World.

The Irish presence and overt actions changed the landscape physically, socially, and culturally. In this matrix we can clearly view the divide between first and second nature. The new migrants found a landscape marked by human presence, since it already bore the comparatively benign marks of anthropogenic burning as well as hunting and gathering activities. The first nature that the Europeans encountered was, in an ironic way, a ready-made location for settlement, with hunting trails already blazed through the forest and trees burned away, creating clearings near rivers and streams. When the newcomers stood at the threshold of the forest in the first half of the nineteenth century, their second nature vision was one of creating homes and a secure community on the forest frontier.

Ecological Locales: Ireland

In order to understand the relationship of the two primary Irish groups with the Trent Valley, we must first uncover their connection to their sending environment. The first group was from County Cavan, part of the Ulster Plantation; the second, from County Cork in southern Ireland. Although Ireland and Upper Canada were quite different geographically, both groups possessed skills, honed from working their native lands, that enhanced adaptation to the New World.

The landscape from which both Irish groups came is prominently marked by glacial features. As geographer T. W. Freeman writes, "[I]t is not remarkable that the two words given to science by the Irish language are 'drumlin' and 'esker.'"[2] Indeed, the structure and land forms of County Cavan are similar to those of the Scottish Highlands. A Caledonia mass known as the Newry Axis gives rise to *drumlin* ("little hill") swarms from the coast of County Down in the northeast to Cavan.[3] The lowlands of County Cork are also rocky, especially the Blackwater Valley, which consists of low limestone ridges, the product of an icesheet that deposited sand and gravel drifts throughout the area.[4] The environment lends itself more easily to transhumance than to diversified agriculture.

The climate and weather also influence land-use regimes. With an average annual rainfall of between thirty and sixty inches annually, and many frost-free months (up to ten in Cavan, and near total in Cork), grassland stock raising is preferable to arable farming in both counties. Indeed, in Cork it is possible to leave a herd of cattle in the field for almost the entire year. In Cavan, cropping usually consists of oats, potatoes, and flax, with very little wheat or barley. Further south, where exotic sub-tropical tree species are known to thrive, similar cropping predominates.[5]

The late eighteenth and early nineteenth century period was the golden age of agriculture, especially in County Cork. The increasing demand for tillage product, instead of stock-rearing and dairy product, during this period prompted landlords and small farmers to expand their activities. The result was an upswing in land reclamation, especially of the former "wastelands" such as bogs or coastal zones, or under-utilized upland foothills, which had been used for grazing.[6]

Human Cultures: Ireland

The ability to adapt to the land was the common denominator between the Irish Protestants and Catholics who eventually settled the Trent Valley. In Ireland, however, their experiences were socially and culturally different.

Protestants from County Cavan, responding to an economic downturn, relocated to Upper Canada in 1817. They did so under the auspices of the British consul at New York, James Buchanan, who financially assisted these immigrants' passage to Upper Canada. They brought with them their place namesake, which carries some significance. In the Old World during the early seventeenth century, the area was part of the Ulster Plantation, one of six counties in the north that became the early testing ground for the anglicanization of Ireland.[7]

The overwhelming majority of the Irish Catholics destined for the Trent Valley originated from the rural Blackwater Valley region of County Cork. The county's economy had long been anchored by its capital and major port city Cork, which had flourished during the era of the Napoleonic Wars as a provision and recruitment centre for the British army and navy. Yet the county as a whole was rural and was characterized by two principal classes of agricultural labourers: bounded and unbounded. Bounded labourers enjoyed a higher degree of security and were less susceptible to seasonal fluctuations than were the casually employed, unbounded labourers. Both groups rented land, and as was common in pre-famine Ireland, employed a land subdivision strategy by which food could be grown and a plot given as inheritance.[8]

However, following the wars' end in 1815, County Cork's economy suffered. Demand for agricultural products subsided, farms began employing fewer bounded and unbounded labourers, and the labour supply increased further as thousands of men returned from military service. Demand for land among the rural poor was a fact of life between 1780 and 1815. Wasteland reclamation occurred throughout Ireland during this period, but in Cork it was practiced at an increased rate in order to keep pace with the demand for arable land due to subdivision. A famine during 1821 – 22 further imperilled the peasants' situation, and riots broke out in the countryside.[9]

The county's situation was stressful for both the rural tenants and for their landlords. Thus when Robert J. Wilmot Horton, the British Under Secretary of State for Colonial Affairs, proposed state-assisted immigration from southern Ireland to Upper Canada the idea found many supporters. In a time of cutting-edge Malthusian theory, some intellectuals and administrators expressed fear that British Isles resources would be outstripped by an increasing population, a situation exacerbated by swelling working-class ranks following the end of the Napoleonic Wars.[10] In response, by the late 1810s, men such as Horton began experimenting with strategies of depopulating impoverished colonies like Ireland and transporting those peasants to newer and more sparsely populated colonies such as Upper Canada. This plan was successfully tested in 1815 and again in 1823, when, respectively, Scottish and Irish Catholic communities were transported to the province.[11]

Peter Robinson, a member of the Legislative Assembly of Upper Canada and an honorary member of the Executive Council, had acted as superintendent for the 1823 assisted immigration of nearly 600 Irish Catholics to the Rideau District. When commissioning that transport, he was deluged with applications. So strong was the outpouring of emigration sentiment that in May 1825 a second passage was arranged from Cork consisting of 2,024 agricultural and casual labourers and their families.[12] These immigrants were so grateful to Robinson that they named the Trent Valley's major commercial centre, Peterborough, after him.

Robinson's selection criteria came directly from Horton and were quite specific. He was to draw overwhelmingly from Cork's Catholic population (the religious group that contained most of the rural poor); no applicant should have means to pay their way; no single men were allowed: only families headed by able-bodied men under age forty-five with no more than three children under age fourteen. Local landlords often sponsored the applicants.

Robert King, the Earl of Kingston, lauded the benefits of the immigration strategy. The project had merit because it siphoned off Ireland's increasingly distressed farmers. The changing land-tenure structure, he believed, was further marginalizing the small farmers. "The very small farmers here," King wrote, "can not now get lands and they will make the best settlers. And if left [here] without occupation will turn into bad subjects, and if they go to Canada will cultivate the waste lands there as they carry with them habits of Industry and will be useful members of Society."[13]

Making a Home Place:
Ecological Locales and Human Cultures of the Trent Valley

In adapting to the New World, Irish immigrants drew upon three fonts. First, they applied land clearance skills that they had gleaned from their homeland. Second, they observed and learned from the Mississauga, especially with regard to burning and fishing. And third, as labourers for families like the Traills, they learned forest clearance techniques while earning hard cash. Their motivation, whether Protestant or Catholic, was to create a new home place in the Trent Valley. The largely Protestant township of Cavan was settled after 1817. Catholics predominated in the townships where they were located in 1825, namely Douro, Smith, Otonabee, Emily, Ennismore, Asphodel, and Ops.

Forest Uses, Forest Clearance, and the Potash Frontier

The Trent Valley contained some of the province's best available pine timber, which was mixed among other softwoods and hardwoods. The Mississauga's burning of the forest produced ash rich in alkaline nutrients; corn and domestic grasses thrived under such conditions.[14] Burning also sped up the regeneration of wild grasses, attracting wildlife closer to the native encampments and saving hunters from excursions into the forest. Corridors for travelling throughout the watershed system were also blazed with prescribed burning. The first Anglo-Celtic settlers followed the natives' example, and some of the area's original roads were created through burning.

While forest clearing techniques were learned from the Mississauga, the Irish possessed some knowledge of fire as an agricultural tool. Among Irish farmers, burning had facilitated the corn harvest prior to the eighteenth century agricultural

Figure 4: Anne Langton, *Peterborough from White's Tavern*, [c. 1837 – 38]; Langton Family Papers; F 1077-8-1-4; Archives of Ontario; AO 2400.

revolution in farming techniques. Fire was useful in at least three ways to secure corn seeds. A farmer might extract seed by burning the straw and husk, leaving the grains scorched but undamaged. Second, the entire sheaf might be lit; once extinguished, the parched seed was separated from the ashes. The removed corn ears might also be burned in a heap, and the seed taken.[15]

In North America the two most common methods for clearing forest were chopping and girdling. Chopping simply involved swings of an axe: girdling was even less labour intensive although it involved a long wait. A woodsman slashed into the bark all around the base of the tree, thus cutting off the flow of sap and causing the leaves and eventually the branches to fall. With shade no longer protecting the moist forest floor, the soil would dry sufficiently so that crops could be planted. Girdling was practised in the northeastern states and eastern Canada, though to a lesser extent than chopping. In these regions, cut trees could be readily sold as logs, lumber, fuel, or ships' masts and were too valuable as sources of income to merely let rot in a clearing.[16]

Many Irish arrived with experience in land reclamation. Those who had had little contact with such activity, however, quickly overcame this deficiency and acquired knowledge by hiring themselves out to more established valley inhabitants. Samuel Strickland, for instance, related his experience of May 1826. He had arrived the previous year and had recently purchased wild land on the banks of the Otonabee River, within a mile of Peterborough. Faced with the unenviable task of clearing he "very foolishly hired two Irish emigrants who had not been longer in Canada than myself, and of course knew nothing either of chopping, logging, fencing, or, indeed, any work belonging to the country. The consequence of this imprudence was," he lamented, "was that the first ten acres I cleared cost me nearly 5 [pounds] an acre—at least 2 [pounds] more than it should have done."[17]

Strickland's loss was the hired Irishmen's gain. They used that experience on their own lands and by 1830 had etched out a settlement on the forest frontier. According to a Kingston newspaper, "[F]rom scenes where they were lingering under distressing despondency and gloomy despair, [they move] to those, where they now breathe the air of comfort, and comparative ease, and look forward with a cheering certainty to approaching independence."[18]

The Irish appear to have used subsistence strategies effectively within the first decade of Trent Valley settlement. Moreover, the Irish Catholics' skills were greatly relied upon by new Anglo migrants to the area. Catharine Parr Traill, sister to Samuel Strickland, recounts that the hiring out of (usually Irish Catholic) labour was common and beneficial to both parties. Occasionally the Traills employed Irish domestics; during their first winter in Upper Canada (1832), Catharine's husband Thomas let out ten acres of their wooded land to some Irish choppers. The men built a winter shanty on the lot and received fourteen dollars per acre for chopping, burning, and fencing that quantity.[19] The following spring they hired an Irish boy to collect sap and boil it down into sugar.[20]

Indeed, maple sugar served the newcomers both as a ready sweetener for their diet and as a marketable product. Sugaring, too, was a technique that they and other Europeans learned from the Aboriginals. The Hurons and other Great Lakes natives demonstrated the process of tapping maple trees each spring, and settlers in New France and later British North America practised the technique quite profitably.[21] This forest by-product was found by mid-century among all of the townships under investigation here, while curiously another forest product, apple cider, was produced only among those pioneers at Cavan (433 gallons in 1851), Monaghan (600 gallons), Ops (10 gallons), and Emily (700 gallons).[22]

Figure 5: Anne Langton, *End view of John's House*, [c. 1837 – 38]; Langton Family Papers; F 1077-8-1-4; Archives of Ontario; AO 2585.

Chopping was the Trent pioneers' first step toward becoming farmers. In his travels through the valley, Englishman Basil Hall found two Irish-Catholic families who had moved closer to that goal. Between November 1825 and July 1827, one family of eleven persons had cleared twenty-six acres, most of which were under cultivation. The other family of two men and their father had chopped more than twenty acres and had a crop of wheat, oats, Indian corn, potatoes, and turnips as the harvest season approached.[23]

After chopping, underbrushing was the important next step. The settler cut away the undergrowth and smaller trees, then piled the debris so that it could be burned. The remaining larger trees were then felled in the fall or early spring, and the best logs were selected for home construction, trade, or other uses. If the remaining trunks were then burned during a dry period, they produced a natural enrichment for the soil. The residue, commonly known as potash, was important not only as a fertilizer but also as an exchange commodity.[24]

Knowledge of this process helps to explain the otherwise prosaic actions of Bobcaygeon farmer and miller Thomas Need in November 1833. He recorded:

November [1]. The men were now employed in logging and burning, a dirty, disagreeable work, at which I was often obliged to assist.

3. To day [*sic*] I bought some seed wheat at a dollar per bushel, and also a yoke for oxen, for which I gave 20 [pounds], not however without some misgivings on the score of prudence, having only secured a very small stock of hay from the Beaver meadow ...

5. The labourers finished sowing wheat, and called me to admire the appearance of the clearing. I believe it was very credable [*sic*] for the Bush, but I was scarcely yet naturalized enough not to take offence at the black unsightly stumps still remaining.[25]

Clearing the forest was an arduous task, but one that needed to be undertaken only once. The soil was altered by the burning and thus rendered suitable for the first planting. The stumps to which Need referred were often left in the ground; grubbing, or removing them was a difficult task, and farmers had other more pressing duties to attend to.

Often clearing was approached communally. Logging bees were organized whereby farmers would aid their neighbours. Traill recalled a summer 1833 gathering when a number of the surrounding settlers brought oxen to their lot. The felled logs were arranged into piles and set ablaze. Such bees were important for community building from both an economic and social standpoint and signifies the interdependence of frontier and rural life.

It has been suggested that in the interim period between clearing, burning, and planting, a local economy emerged in rural communities throughout North America. Described as a by-product of the clearing process, potash and pearlash manufacture became the first cash crop of nineteenth-century farmers.[26] The practice had its roots in the colonial export economy and was a feature of the St. Lawrence River–Lake Ontario corridor. In the early 1820s, potash rivalled wheat as an export commodity, and high prices attracted farmers and merchants throughout Upper Canada. Potash was commonly shipped in barrels usually weighing between four and six hundredweight. The more refined pearlash, weighing about three hundredweight was shipped in smaller barrels. [27] Local historian Harold Pammett notes that by 1825 Emily township pioneers were trading potash and grain for oxen and cattle.[28]

Hardwood ashes could be transformed into potash and its refined derivative, pearlash. By repeatedly pouring boiling water over the ashes in large kilns, the ashes were turned to liquid lye. Lye was an alkali essential to the manufacture of soap, glass, tannin, bleach, medicines, and other chemical products that use potassium carbonate compounds.[29] The principal extraction from boiled ashes, lye was a household product that, when combined with kitchen greases or animal entrails, produced soap.[30] Large cast-iron, bowl-shaped kettles were filled with the wood ashes and water and then boiled. Over a two- or three-day period the lye was collected in a receiving trough. Once the contents of the kettle became thicker and darker, the lye filtering ceased, but the boiling of ashes required constant supervision. Women stirred the pots with large wooden spoons. After about three or four days of attention, the potash was a dark red colour and resembled molten metal. Once cooled it became a solid mass and was barrelled and sold to merchants for upwards of thirty dollars per barrel.[31]

The activity fit well with the rhythms of rural life. As pioneers cleared the forest cover, a ready market was available for their settlement by-product. The potash and pearlash trade reached its zenith during the 1830s, though it continued throughout the Trent Valley for at least another decade.[32] During the early 1830s Cobourg merchant Charles Clarke paid cash for agricultural produce and potash.[33] By the 1840s, however, Peterborough's William Eastland and Henry Easton reverted to a more rudimentary exchange economy and took all "country produce" including potash *as* cash.[34]

Agriculture and Livestock Pasturing

Agricultural activity, coupled with livestock raising, continued to rank among the chief alterations imposed by the newcomers upon the Trent Valley landscape. Indeed, within one generation valley farmers were not only subsisting on their produce but also seeking to enhance their livelihoods by supporting plans for a canal, which would better connect them to markets both within and beyond their region.

Like other settler groups throughout North America, those of the Trent Valley had to craft adaptive strategies in order to survive. These settlers had a point of reference. The drumlin and esker swarms of the sending country were similar in terrain and in agricultural potential to the drumlins that characterized the landscape around Rice Lake and the Otonabee River. The Irish, in other words, possessed some of the technical expertise with which to control, and hence adapt to, their new surroundings.

Reclamation both in Ireland and in Upper Canada was a non-capital-intensive endeavour; it could be undertaken primarily with tools available to the Irish in

Table 3.1:
Potato Harvest by Township, Trent Valley Irish Catholics, 1825–26

Township	Cleared Acres	Potatoes in Bushels	Bushels per Acre
Douro	245.5	8,251	33.6
Smith	113.25	4,800	42.4
Otonabee	186.0	10,500	56.5
Emily	351.5	22,200	63.2
Ennismore	195.0	8,900	45.6
Asphodel	173.0	9,150	52.9
Marmora	35.0	1,198	34.2
Ops	12.0	800	66.7
Total	1,311.25	65,799	50.2

Source: Archives of Ontario, F61, Peter Robinson Papers, "Settlement Projects of Peter Robinson, 1823 – 1826" [typescript, Howard T. Pammett, Peterborough Public Library, 1933] (MS 524).

both settings. Spades were among the implements distributed to the Irish Catholic immigrants when they arrived in 1825. More importantly though, the Irish colonization of the foothills and the drumlins was facilitated because they were armed with their own "ecological ally": the potato. This tuber was adaptable to both Irish and Canadian cool climates and to moist soil ecosystems.[35] Each of the eight townships recorded sizeable potato crops during the first year (see Table 3.1).

This most basic case of the potato crop provides both a preface to a discussion of Irish adaptability to the Trent Valley landscape and a trans-Atlantic link between rural life in Ireland and Upper Canada. Certainly some elements of life in the New World prompted adjustments within the Irish community, yet these immigrants were not entering the wilderness without some measure of experience.

The statistical data presents a glimpse of progress during the first generation of Anglo-Celtic settlement, a progress often defined in terms of agricultural production. As alluded to in Tables 3.3 and 3.4, crop diversification was common among both of the transplanted Irish groups. Potatoes and turnips were the most common and basic of the crops at the outset. However, by mid-century barley and hops (ingredients used in brewing beer), rye (used in bread making or sold to distilleries),

Table 3.2:
Agricultural Product and Purchases, Trent Valley Irish Catholics, 1825 – 26

Number located	437
Acres cleared	1,311.25
Potatoes (bushels)	65,799
Turnips (bushels)	24,173
Indian corn (bushels)	10,440.50
Wheat (bushels)	338.50
Maple sugar (lbs.)	9,067
Oxen purchased	38
Cows purchased	69
Hogs purchased	152

Source: Archives of Ontario, F61, Peter Robinson Papers, "Settlement Projects of Peter Robinson, 1823 – 1826" [typescript, Howard T. Pammett, Peterborough Public Library, 1933] (MS 524). Townships of Douro, Smith, Otonabee, Emily, Ennismore, Asphodel, Marmora, and Ops.

peas (preferred by most as a means for fattening hogs), oats (used as feed for horses), carrots, beans, hay, and flax were also important.[36]

As was the case in most Upper Canadian agricultural communities, wheat was the principal crop of Trent Valley farmers. As historian Robert Leslie Jones notes, "[w]heat was by no means the only crop grown by the wheat farmer; it was simply the one which he sold ... [w]hile peas, Indian corn, oats and barley had a large aggregate production, they were primarily intended to be consumed on the farm."[37] For Upper Canadian farmers, wheat was a chief import and could also be locally milled to meet their own needs.

The advent of Red Fife wheat during the late 1840s complemented the area's agricultural product. Early frosts, wheat rust, and other diseases annually wreaked havoc on farmers' crops. In response to these threats, Otonabee Township's David Fife began experimenting with northern European wheat strains in an effort to overcome the North American impediments. In 1820, at age seven, Fife had come from Scotland to Upper Canada with his parents. By the early 1840s he was practising mixed farming on his own 200 acres. Fife received a sample of Galician wheat from a friend in Scotland in 1842, which he grew successfully. In time, the Red Fife wheat surpassed the local strains in durability, and by 1848 it was distributed

Table 3.3:
Agricultural Product and Purchases of Cavan Township Irish Protestants between 1826 and 1851 – 52

PRODUCT/PURCHASE	1826	1851 – 52
Number of persons assessed	255	472
Acres cleared/under cultivation	25,101	26,056
Potatoes (bushels)	n/a	19,427
Turnips (bushels)	n/a	23,522
Indian corn (bushels)	n/a	1,670
Wheat (bushels)	n/a	108,319
Maple sugar (lbs.)	n/a	17,216
Oxen	272	826
Milch cows	416	1,516
Young cattle	198	1,627
Hogs	n/a	6,070

Sources: Archives of Ontario, MS 180,"Aggregate Census and Assessment Returns for Upper Canada, 1825 – 1849,""A General Account of the Rateable Property in the District of Newcastle for the Year Ending Upon the First Monday in the Month of January, 1827"; Province of Canada, *Census Report of the Canadas*, 1851 – 2, vol. 2, no. 6,"Upper Canada Return of Agricultural Produce for 1851 – 2" (Quebec: King's Printer, 1855), 8 – 11.

to other area farmers. By 1851 its popularity had spread to some mid-western American states, and later in the century this seed aided Canadian Prairie farmers as they formed agricultural communities.[38]

Despite such innovations, some agricultural continuity with Aboriginal practices could still be found. Indian corn, a species cultivated by the Mississauga and adopted by Anglo-Celtic pioneers, remained a crop on Trent Valley farms in the 1850s, though in vastly reduced quantities; most farmers chose instead to devote acreage to wheat. The best feature of corn was the ease with which it could be grown. Traill, for example, reports, "The cultivation of Indian corn on newly cleared lands is very easy, and attended with but little labour; on old farms it requires more. The earth is just raised with a broad hoe, and three or four corns dropped in with a pumpkin-seed, in about every third or fourth hole."[39] The corn was preferred by, as Traill put

Table 3.4:
Agricultural Product and Purchases of the Townships in which Irish Catholics Were Located between 1825 – 26 and 1851 – 52

Product/Purchase	1825 – 26	1851 – 52*
Number located/occupiers	437	2,004
Acres cleared/under cultivation	1,311.25	78,998
Potatoes (bushels)	65,799	107,619
Turnips (bushels)	24,173	80,507
Indian corn (bushels)	10,440.50	4,109
Wheat (bushels)	338.50	318,006
Maple sugar (lbs.)	9,067	31,968
Oxen purchased	38	4,561
Cows purchased	69	6,150
Hogs purchased	152	12,733

* Marmora township has been excluded from this category. Although Marmora received Irish Catholics in 1825, few chose to permanently locate there.

Sources: Archives of Ontario, F61, Peter Robinson Papers, "Settlement Projects of Peter Robinson, 1823 – 1826" [typescript, Howard T. Pammett, Peterborough Public Library, 1933] (MS 524). Townships of Douro, Smith, Otonabee, Emily, Ennismore, Asphodel, Marmora, and Ops; Province of Canada, *Census Report of the Canadas*, 1851 – 2, vol. 2, no. 6, "Upper Canada Return of Agricultural Produce for 1851 – 2" (Quebec: King's Printer, 1855), 40 – 3, 48 – 51.

it, "many enemies," including bears, racoons, mice, and birds, but especially cattle. In fact, "breachy" cattle had been know to toss down a fence to get at the crop.[40] Traill's observation about hungry cattle, disrespectful of defined bounds of property, leads us to a kindred topic below, that of "bounding" the land.

Thus, as with forest clearance, agriculture began to move the Trent environment further from its character during the period of Mississauga dominance. To be certain, some similarities between Mississauga and Anglo-Celtic subsistence regimes were evident. For instance, both groups were at the mercy of the North American climate and seasonal changes, which included inhospitable winters. But the differences were more plentiful and more striking, and the examples from newcomer livestock raising and production help bear this out. As William Cronon writes of the contact period in New England, "what made Indian and European subsistence cycles seem so different from one another had less to do with their use of plants than their use of animals."[41]

Livestock Rearing and Pastoralism

The Irish were well connected to the land in both the sending and receiving contexts. No example better illustrates this than the case of livestock rearing, pasturing, and production.

In both contexts, wealth was measured in heads of cattle. In the drumlin-dotted, well-hydrated County Cavan and in County Cork with its mild climate and grass-abundant coastal environment, transhumance and dairying anchored the local economy.[42] It seems that the transfer of this practice to the New World was one not always of convenience but of necessity on the drumlin landscapes that could so easily circumscribe agriculture; however, the newcomers recognized that although the land was sometimes difficult to farm, it afforded other uses. For instance, by 1851 – 52, more than one-third of cultivated land was given over to pasturing in four townships: Cavan (43%), Emily (40%), Douro (38%), and Ops (36%); in another three, more than one-quarter of the land was being used in the same way: Asphodel (32%), Smith (27%), and Otonabee(27%).[43] Adaptation to the Upper Canadian environment was an ongoing challenge, but one that the Irish met with success during their first twenty-five years in the New World.

Certain types of stock-rearing, such as hogs, were popular in the Trent Valley.[44] This precipitated the bounding of the land. As Traill recalls, pigs "are great plagues on a newly cleared farm.... If they run loose they are a terrible annoyance to both your own crops and your neighbours if you happen to be within half a mile of one; for though you may fence out cattle you cannot pigs."[45] Fencing was perhaps the necessary and logical sequence in the process that began with the surveying of township plots even before the settlers arrived. In Upper Canada, as in New England, the township model was in place, usually demarcating space in 100-yard, squared parcels.

Cronon notes that livestock in New England was a new addition to the landscape and quite foreign to native peoples.[46] Draft animals, for instance cows, were producers for the settlers: milk, cheese, and butter were essential not only to the diet of farmers and were key elements in their commercial lives. By mid-century, all the valley townships in this study were producing butter and cheese and barrelling beef and pork.[47] Oxen were key to forest clearing operations and to ploughing fields, which helps to explain why in 1827 Cavan settlers, even after having lived in the area for a decade, possessed many more oxen (272) than horses (32).[48] Horses were used for personal transportation, which tended to encourage the speedy completion of roads in the young province.

In addition to being a new species on the Trent Valley landscape, sheep, like cows, were important to the family economy as they produced wool that could be made into clothing (each of the townships under investigation produced flannel) or sold at market.[49] This helps to explain why a bounty on wolves, a predator of sheep, was an early feature in the Trent Valley.

The first generation of settlers cared little for maintaining a balance between themselves and animal populations. They could not see the benefit of maintaining an equilibrium between predators (wolves) and varmints (chipmunks and squirrels), both of which confounded agriculture. Wolves were targeted by livestock farmers within the first decade of settlement. Although the front townships were clear of the predators, rear lot farmers' livelihoods were threatened. By early spring 1835 as many as fifty sheep had been killed, and local men called for bounties on wolves to rid the countryside of the baneful creatures.[50] Ironically, only eight months earlier another farmer, one whose crops were being devoured by chipmunks and squirrels, had advocated a bounty on these varmints.[51] Obviously the wolves might have kept the chipmunk and squirrel populations in check, yet farmers saw no utility in protecting either species.

All of these new uses of nature also implied human "ownership" of it. This was an obvious departure from the Mississauga conception of the bioregional community. For Aboriginal peoples, deer in the forest and fish in the rivers and lakes could not be owned, whereas Anglo-Celtic settlers' cows and sheep, bounded within defined spaces, could be. Thus when the Mississauga petitioned for protection of game and fish, they did so with the understanding that these resources were intended for widespread use. However, when settlers acted to rid wolves from grazing areas, they clearly sought to protect their personal investments.

Fishing and Hunting

Although few references exist in the historical record, newcomers, like the Mississauga in their midst, must have supplemented their diet with fish and game.

Figure 6: Anne Langton, *River at Peterborough,* [c. 1837 – 38];
Langton Family Papers; F 1077-8-1-4;
Archives of Ontario; AO 3171.

We know from the accounts of Thomas Carr, Catharine Parr Traill, and Thomas Need that the Mississauga caught and sold fish. The Anglo-Celtic settlers, however, appeared not to have engaged in fishing to a large extent. Moreover, by mid-century, despite their proximity to rivers and lakes, only three townships recorded quantities of cured fish among their annual produce: Emily (8 barrels), Ennismore (6 barrels), and Ops (2 barrels).[52]

We can speculate that hunting, too, was an activity that added proteins to the newcomers' diet, although references are remarkably sparse in the early literature. Levi Payne, an Englishman in Dummer Township, lauded the abundance of game in his area, noting that partridges, deer, woodcocks, snipes, ducks, geese, foxes, beavers, and bears were plentiful. As well, Payne, corresponding with his parents in 1831, wrote almost with glee that a veritable "open season" existed on wolves; "for killing a wolf we get 1 pound, 5 shillings, we take the head to the governor and receive the money."[53]

Mills

Other forms of claiming the land also existed in the Trent Valley. Grist and sawmills were two of the earliest expressions of the pre-industrial age in Upper Canada. The analysis will become more specific and detailed in the next chapter, but here we will begin to investigate yet another means by which European newcomers shaped the landscape to meet their needs. The early townships of the valley—Cavan, Monaghan, and Asphodel—had at least one sawmill each in 1826;[54] by 1836 that number had increased to include at least one sawmill each in Emily, Otonabee, Douro, Dummer, Smith, Ops, Fenelon, and Verulam.[55] This trend remained constant until the mid-1800s, with Smith Township alone possessing six sawmills by 1852.[56]

Grist- and sawmills were significant structures in frontier communities, holding out the promise of moving communities toward greater independence. Wheat could be made into flour for home use; cut logs could be reproduced as sawn lumber both for home construction and transshipment to distance markets. Early on, Anglo-Celtic settlers were cognizant of the need for a millsite. Emily Township's new Irish settlers recognized that their settlement's success depended upon the construction of a mill, the nearest one being twenty-five miles away. Seventy-four men and women petitioned Lieutenant-Governor Sir Peregrine Maitland saying that although they were "truly grateful" for the many favours bestowed upon them by the government, they were unhappy about the "'impossibility' of making a living for themselves and families without a Mill."[57] Likewise, Traill asserts, "how important these [mill] improvements are, and what effect they have in raising the spirits of the emigrant, besides enhancing the value of his property in no trifling degree. We have already experienced the benefit of being near the saw-mill, as it not only enables us to build at a small expense, but enables us to exchange logs for sawn lumber."[58] Sawmills in particular complemented the work of forest clearing; Traill continues, "The great pine-trees which, under other circumstances, would be an encumbrance and drawback to clearing the land, prove a most profitable crop when cleared off in the form of saw-logs, which is easily done where they are near the water."[59]

Although mills performed a necessary function, their construction altered the landscape. For example, the mills owned by Zacheus Burnham in the northernmost portion of Dummer Township and by Dr. John Gilchrist at Keene near Rice Lake were both powered by the flow of the Indian River. In 1840 Burnham and Gilchrist realized that water flow from river's source at Sun Lake often failed to provide adequate power for the mill during the autumn and winter. They therefore used dynamite to deepen the course from Stoney Lake to Sun Lake, raising the water levels along the Indian River and increasing the supply of water for their mill concerns.

All of this was done without the permission of the Crown Lands Department, which had a claim on the land. Thus two private citizens acted to change the flow of water to benefit their business needs in order that, as Burnham explained, "the Inhabitants can have their grinding and sawing done without delay and disappointment."[60] Evidently, the alterations to the watercourse were successful, as by 1858 the Keene facility was sawing 2,000 feet of lumber and milling 250 bushels of wheat per day.[61] The actions of Burnham and Gilchrist, however, were slight compared to the major project that local boosters and the province undertook in constructing the Trent Canal.

Roads and Canals

Until this point, communication and transportation from the back townships of the Trent Valley to those bordering Lake Ontario (such as Port Hope and Cobourg) had relied tenuously on poor roads.[62] The province had underwritten canal projects in the Rideau District and at Welland, and within the first generation of settlement, local farmers and merchants were clamouring for a similar commitment to the Trent watershed. Support for the project was first voiced in 1833 and built steadily, culminating in the start of construction during 1836. Canal supporters throughout the area lamented the limited access that kept farm produce and valuable timber from reaching larger North American markets.[63] In 1836 "A Veteran Pioneer" summarized the problem and urged his "Brother Backwoodsmen" to petition for the canal's construction. "There *is* a hole,—a wide but not irreparable hole,—in your coats," he warned. "[Y]ou possess a section of the Province, fertile in the production of all kinds of grain, especially wheat; abounding in the finest timber, with rich iron ore; and watered by a chain of lakes and rivers, which, at a moderate expense in improving them, hold forth to you and your offspring, the inestimable advantages of an inland navigation ... [A]t present you are shut up and locked in; 'cabined, cribbed, and confined.'"[64] As if to rightfully situate the valley among more established communities of the province, he continued, "[Y]ou raise all kinds of grain, but you cannot carry it to market, except during the middle of the winter. Whilst your brother farmers in the front townships, with land inferior to yours, are selling *their* wheat ... and hogs [at higher prices]. Think of the advantages you possess, and of the profits you are yearly losing.... *The upper and lower parts of the Province have been benefitted by canals, but ours has been totally neglected."*[65]

By mid-century, the Trent Valley was changing rapidly. Europeans, who were replacing Aboriginal peoples as the dominant cultural group, had blazed trails through the forest, created roads, planted crops, and erected fences to bound the land. A canal was under construction, and lumbering would soon become a mainstay of the local economy as a trade nexus took shape. All these developments occurred within a rough twenty-five year period.

The divide between first nature and second nature had indeed widened. The Mississauga certainly had altered the landscape to supply their needs. This was moderate, however, compared with the changes inspired by European settlement with its manifold uses of nature. As well, the diverse social fabric gave rise to differing conceptions of "home place" within the Trent Valley.

The Home Place

Thus far we have been concerned with the transfer of skills, the acquisition of local knowledge, and adaptation to a new environment. We have not yet considered the social fabric woven from the process of resettling the land. Here we take up the home place, which, as Donna Birdwell-Pheasant notes, was of utmost importance to peasant families in Europe. However important land was to Irish families, she suggests, it should be understood as being "conceptually embedded in a vital nexus of rural life that was governed by two deeply consonant ideas—the idea of family and the idea of place."[66] What kind of home, then, did the Anglo-Celtic migrants seek, and how did they proceed to establish it?

By 1833 the village of Peterborough was declared the "chief emporium of the back settlements" and served as the centre for the district's hinterland. Here farmers found a ready market for their surplus wheat, corn, and other product, although larger market centres, such as Montreal, were kept beyond reach for want of a direct trade route to the Bay of Quinte. Peterborough retained the character of a frontier town: pine stumps still littered the ground, the roads were haphazardly planned, and the homes were crudely constructed. Nonetheless, one correspondent saw nothing but opportunity when he surveyed the watercourses that made the town the linchpin of the Kawarthas–Trent River system. It was also "the most loyal village in Upper Canada ... an excellent example of what a British settlement is sure to be under good management."[67]

The sense of community helped to facilitate the migrants' acclimation to their new surroundings. One study of the valley's Irish Catholics has underscored the close proximity of immigrant landholdings. Social institutions such as the nuclear family, kinship, and antecedent communities influenced decisions to locate within the area. Alan G. Brunger surveyed the record of initial Catholic families and found a high degree of continuity between the rural communities in Ireland and those which emerged in the original Trent Valley. For example, all ten families from Listowel, Kerry located in Ennismore; eighteen of the twenty-six families from Brigtown, Cork located in Douro; twelve of seventeen Mallow, Cork families and eleven of seventeen Kilworth, Cork families settled in Emily. In these cases communities were "transferred" to Upper Canada.[68] In addition, in the eight townships members of nuclear and expanded families tended to settle close to each other. Initially seventy-nine of the 115 families chose contiguous locations, while fourteen family members lived less than one mile away from one another.[69] Such proximity to neighbours, both old and new, provided a social support network that helped new immigrants, as in cases when family members aided one another with food, support, and shelter.[70]

However, while certain continuities can be discerned among the new immigrants, John Mannion found that, similar to other Upper Canadians, the Irish Catholics were also influenced by the commercial agriculture nexus. In Ireland, land scarcity encouraged farmers to intensively work the arable areas; however, Upper Canada saw the converse: comparatively vast areas, but few labourers. Throughout the Trent Valley "extensive" farming prevailed and cereals such as wheat were cropped. Irish newcomers adapted to this system; soon they began shifting crops and fallowing their fields so as to reduce rotations. For instance, after the first year of settlement, Ennismore pioneers worked an average of three acres per family. That rate increased in 1830 to five acres, and to twelve by 1839. This new method better suited the land-labour dynamic and enabled pioneers to spend less time farming and instead to pay more attention to forest clearing.[71]

Occupational pluralism was, for some newcomers, an economic reality. The traveller Basil Hall explained that it was common for many newly arrived emigrants to gradually move toward economic independence by way of seasonal employment. Such jobs could entail canal work for adult men or, in the case of boys and girls, domestic servitude in neighbouring villages or towns such as Cobourg. Family members might engage such opportunities for the first few years, long enough to purchase livestock or farm implements.[72] In time, he explained, "[t]he progress towards independence ... is very rapid when industry is applied to the untouched soil of that country, and the parents are enabled gradually to withdraw their girls and boys from a description of [wage-earning] service" in adjacent towns.[73] Hall believed that given this prevailing drive toward independence, it was easy to explain why during the first year of settlement by a "fresh batch" of immigrants, domestic labour was plentiful in large towns like York or Cobourg. However, as the emigrants succeeded "in establishing themselves, and thus acquire independence, so the difficulty of obtaining servants increases."[74] What Hall failed to realize, however, was that many Cork immigrants were farm and domestic labourers in Ireland and were drawing upon past experience. By selling their labour or that of their children for a short period, they were able to put the earned wages toward the purchase of livestock or perhaps a better home.

The newcomers were thankful for the new opportunities made available to them in Upper Canada. This was especially true among the second wave of settlers, Catholics from County Cork, who arrived as part of the assisted immigration orchestrated by Peter Robinson. Asphodel and Douro Townships' inhabitants, for example, made eloquent declarations of thanks and pledges of loyalty to the province and Crown. The Asphodel settlers' 1825 petition is a good example.

"We the ... Irish Emigrants recently brought to this Country by [His Majesty's Government]," they wrote, "...express to your Lordship our grateful sense of the numerous favours we have experienced from [your] kind patronage. For the liberality of a humane and benevolent Sovereign no language can express our gratitude in having removed us from misery and want to a fine and fertile country where we have the certain prospect of obtaining by industry a comfortable competence, and we trust my Lord the report of progress we have already made on our lands will not fall short of your Lordship's expectations, taking into consideration that we have had to contend, in addition to inexperience, with the enemy of all newcomers the fever and ague to a very great extent, notwithstanding which we have been able to provide ample provision to support our families comfortably until we harvest our next crop.... [A]bove all we rejoice that in this happy country we are still under the Government of our truly illustrious Sovereign to whose sacred present Government we beg to express the most unfeigned loyalty and attachment."[75]

Indeed, the importance attached to "loyalty" is significant. As we will see, the attachment to land, its ownership by the newcomers, and the valley as their home place (a place worth defending) became paramount as the bounds of the frontier were pushed back and further "improvements" to the land were undertaken.

Thus by the mid-nineteenth century the resettlement of the Trent Valley was well underway and new relationships with the environment were evident. Anglo-Celtic peoples faced similar challenges of adaptation as had the Mississauga. The result of forest clearance, cropping and pastoralism, and general reconceptualization of nature evinced a different second nature understanding of the Trent Valley as a home *and* place of production. The wedding of the land with the new subsistence regime and the commercial economy may, in fact, be the most apparent of the many changes visited upon the land during this transitional phase. Equipped with the local knowledge gained from adaptation, the Anglo-Celtic groups moved to further shape their environment, sometimes with trepidation, as in the case of Ops Township, or more deliberately, as was the situation along the Bobcaygeon colonization road.

4

Damning the Dam:
Ecology & Community in Ops Township

IDEAS of nature ...," wrote Raymond Williams, "are the projected ideas of men."[1] These words resonate when considering the complex and multilayered ecological history of Ops Township. Much rests upon how people saw the Lake Scugog lowlands and what they envisioned could be created there, thus returning us to the waxing and waning of the first nature versus second nature paradigm.

Ops Township lay west of the most concentrated settlement of Irish Catholics, which began in 1825, and it was located upon poorly drained land. In locating newcomers here, Upper Canadian administrators envisioned a farming community on the banks of a sluggish watercourse. Such a community's material needs could in part be met by a central mill, which, as we saw in Chapter 3, was of great importance to settlers. Anglo-Celtic migrants, for whom little choice land was available in and around the townships more central to Peterborough, sought a home in Ops. The miller who also chose to make Ops his home, William Purdy, ventured into the back country equipped with the technical and business expertise that he had honed for fourteen years in the more established area north of York. Purdy saw economic opportunity, but his naivety and hubris would not only prevent him from advancing this goal, but would create social tension and vast ecological alterations as well.

The home place sought by the Anglo-Celtic settlers was problematic from the start, for as an ecological locale it was swampy and prone to mosquito infestation. It was obviously a difficult place to practice agriculture. Nonetheless, the settlers, many of whom were part of the 1825 state-assisted immigration orchestrated by Peter Robinson, put down roots and despite the less-than-ideal situation created a residence in Ops. This home included a mill, an essential element in realizing material progress and independence on the frontier. Ops people were willing to accommodate the short-term sickness and inconveniences that accompanied the mill's erection. However, when their public right to a home place as loyal subjects of the Crown was threatened, they reacted to the excesses of the outsider, William Purdy.

Between the 1820s and 1840s Upper Canadians reshaped the countryside to accommodate the burgeoning capitalist political economy. Canals became the means by which settlements north of Lakes Ontario and Erie were connected to centres of commerce. The Rideau, the Welland, and the Trent Canals were the centrepieces of such a project underwritten by the provincial government during this period.

An ambitious plan to artificially link Lake Simcoe and Lake Ontario was undertaken in the early 1830s. However, rather than mimic the yeoman work of James T. Angus and chronicle the entire project's history, I am concerned here with the confluence of settlement, capitalism, and state-led economic development as they affected the social and ecological foundations of one community.[2] It was in Ops Township, between 1826 and 1841, that the most profound change accompanying the canal building occurred as the Anglo-Celtic pioneers and one miller struggled to define the environment and foster a community.

"The largest mill-dam in the world"

On 19 October 1833 John Langton, a gentleman farmer from Fenelon Falls, set out to buy potatoes and lumber in neighbouring Ops Township. After an arduous trek through swamp and thick forest, he reached the township centre and "the largest mill-dam in the world," at Purdy's Mills. Langton had not stumbled upon an oasis, but rather a structure that divided the young community. In constructing a fourteen-foot-high dam, William Purdy had altered the relations between the Ops township inhabitants and their surroundings. The mill dam raised the water level of the Scugog River seven feet and made navigation possible for thirty-seven miles south from Purdy's Mills to the river's point of origin at Lake Scugog. Though this pleased some area merchants whose crafts now plied these waters, the price of this success was indeed high for Purdy's agrarian neighbours, as 11,000 acres became flooded as the water rose. Seven area mill sites, one forty miles away, were also inundated, effectively halting Purdy's competitors.[3]

Langton's account serves as a harbinger for the discontent and tragedy that marred this community. Purdy's mills occupied an ambivalent position, and area citizens were divided along the lines of those negatively affected by the flooding and those who benefited from it. While the mills met settlers' demand for lumber and flour, they were also viewed by Ops people as an example of the injurious excesses of a local capitalist. Later still, when contending with malaria originating from the swamps and millpond, the inhabitants would regard Purdy's mill dam as the cause of their illness.

The Natural Setting

THOUGH named after the Greek goddess of plenty, the Ops site ironically had many drawbacks. The land is set within the Schomberg Clay Plains, which extend from Newmarket east to the Scugog area and are characterized by imperfectly drained basins wherein water from the rolling hills collects.[4] The swamp-like conditions that prevailed were best characterized by the Huron word "Scugog" meaning "submerged or flooded lands."[5] The earliest surveyor of the area concurred with this description when he found a diverse hardwood and softwood forest composed of elm, beech, birch, hemlock, cedar, and tamarack set within swamp.[6]

Despite the poor topography, the Crown Lands Department began locating settlers here in 1826. Duncan McDonell led a party from Glengarry in the Eastern District to Ops during the spring and summer months. Unfortunately, shortly after arriving, all were "taken ill with *fever* and *ague* and were forced to abandon the settlement and return to Glengarry."[7]

The fever and ague were, in fact, malaria. Though most often identified as a tropical disease, in the nineteenth century malaria, in its *Plasmodium vivax (P. vivax)* species, was also found in temperate regions like North America. The term "malaria" originates from the Italian *mala* and *aria* ("bad air"), since its cause was once thought to be exposure to swamps, marsh air, or miasma. The disease is actually spread to human populations by anopheline mosquitoes (malaria carriers), which breed *in* the swamps and marshes. When mosquitoes penetrate human skin to draw blood, their salivary secretions enter the bite site and transmit the malaria parasite, which, after a two- to three-week incubation period, attacks its host. In the case of *P. vivax* the incubation period may be prolonged from nine months to one-and-a-half years. Typically the disease is fitful, beginning with a period of chills, then turning to high fever followed by a period of sweating when the patient's temperature drops precipitously. This particular type of malaria is marked by severe illness yet low mortality unless, of course, it is left untreated or inadequately treated. *P. vivax* malaria has a prolonged latent period in temperate regions. For example, a person initially stricken with the disease may encounter a relapse eight to ten weeks later, or at the extreme thirty to forty weeks later. Thus, those ill with malaria in one year may experience a return of symptoms the following year, most often during the spring and summer months.[8]

Malaria is an Old World disease most likely introduced to North America by early European explorers who carried the pathogen in their blood systems. Similar cases of malaria were recorded in other parts of Upper Canada. Rideau Canal workers were severely afflicted in the late 1820s and 1830s, while just south of Georgian Bay, Mono Township pioneers were also stricken. As construction of the Trent Canal progressed in the 1840s, Fenelon and Verulam Townships residents also encountered ague when the government dams at Buckhorn and Bobcaygeon failed and flooding ensued. In the American Midwest, eighty per cent of Illinois's Pike County settlers died from ague in 1825. Pacific Northwest populations of the Chinookan and Kalapuyan were significantly thinned as well.[9]

The only effective measure against malaria was quinine, which attacks the disease-causing parasite. The antidote was available only in select Trent Valley locales during the early 1840s, and its scarce supply might account for the lack of treatment the sick received during the 1820s and 1830s. From Fenelon Falls in November 1842, Anne Langton, sister to John, recorded that "we were obliged to send for a collection of little phials of quinine, which are still in great circulation, and seldom a day passes that one does not appear to be replenished. I wonder how many doses of medicine I have weighed in the last four months! I think almost as much as some village apothecaries."[10]

Second Nature:
Toward a Home Place

Undeterred by the Glengarry group's setback, the provincial government attempted again to promote the township's settlement, and in June 1829 Irish immigrants from the Robinson group were located in Ops.[11] Ostensibly, this was done because the other valley townships had received their full complement of settlers. Crown Lands agents saw the Scugog stream as a potential commercial transportation route within the valley, connecting to the northeast with the larger Kawartha–Trent network.[12] In the interim, however, overland travel from Ops to Peterborough merchants and mills proved difficult and unpredictable. The swampy conditions often mired oxen and wagons, thus complicating food shipments such as flour and pork from the regional centre.[13] It soon became evident to the Crown's agent, Colonel Alexander McDonell, that the new township required a closer gristmill. McDonell wrote to the Commissioner of Crown Lands, Peter Robinson, that "it would be a great loss to the settlement if [the mill site on the Scugog] should not fall into the hands of some capitalist as from its nature it is well worth the attention of such."[14] McDonell even set aside 400 acres for the mill, on lot twenty-one in the fifth and sixth concessions, and believed that the location would act to anchor the town. The lots were intersected northerly by the Scugog stream. At this location the water turned to rapids, the only such point on the waterway, convincing the inhabitants of the site's appropriateness for a mill.[15]

Land clearing and shanty construction continued in and around the township. For the latter activity, McDonell hired two "Canadian" axemen in August 1829.[16] They had been contracted to build twenty to thirty shanties, although construction quickly came to a halt when both men fell ill with ague. Several settlers and McDonell himself were also stricken with fever.[17] The sickness pervaded the Ops community of 127 persons.[18] In fact it became so widespread between the summer and autumn of 1829 that day-to-day tasks such as land clearance and crop plantings were delayed or were impossible to complete. The inhabitants also began to exceed the monthly proscribed food ration which was an attractive and necessary feature of the imperial settlement scheme. By mid-October 1829 the community had become two months overdrawn on its provisions. Knowing the plight of the ill and starving settlers, McDonell wished to extend to them another month's provisions but could not do so without official sanction. He wrote Robinson in York, but due to distance and administrative delays no immediate action was taken on the matter, nor were settlers provided with short-term financial relief.[19]

Frustrated by the delays, the settlers took matters into their own hands. On 25 November 1829 James McNeirney travelled to York and presented Robinson with a petition signed by twenty-three Ops inhabitants demanding that their concerns be addressed. The petitioners intimated that they would abandon the settlement unless immediate action was taken. This ultimatum prompted Robinson to bring the matter before the lieutenant-governor that day. After receiving consent, Robinson instructed McDonell to advance the provisions to the settlers, the cost of which was to be repaid by them within two years. McDonell followed the commissioner's orders, and the Ops families were provided with pork and flour for the 1829 – 30 winter months.[20]

Despite the additional supplies, the symptoms of ague persisted. The swampy conditions that spawned the illness had not subsided, perhaps due in part to the mild winter—by January 1830 no snow had fallen.[21] The unhealthy situation resumed in the summer of 1830. After an August visit to Ops, McDonell wrote to Robinson: "I am in hopes that the fever and ague will be of short duration among the settlers this year, as it already seems to be on the wane, the people in Ops seem to look upon an occasional visit from it as a mere matter of course."[22] The first year in Ops Township was thus a discontented one, with the ague bringing sickness to many of the inhabitants. Chronic food shortages exacerbated by administrative delays only served to anger the distressed settlers.

The events within the township and the actions of its inhabitants demonstrate two points. First, there was a lack of readily accessible and inexpensive provisions such as flour, as well as sawn lumber for home construction. This point was echoed by McDonell in his call for a local mill that could meet the settlers' demand. Second, when distress and frustration escalated to rage, the pioneers would not hesitate to directly confront those in charge.

Second Nature:
The Pre-industrial Context

PRE-INDUSTRIAL mills occupy a paradoxical place in North American history. American historians have rigorously engaged the intersecting themes of property, public rights, and environmental change in which mills proved problematic. In the early Republic the clash between English common law tradition, republicanism, and capitalism gave form and substance to water and property law debates. Public concerns for farming, fishing, or access to water flow were pitted against the economic interests of millers. The law gradually came to accommodate such entrepreneurs at the expense of other riparian users, and the resource itself became an extension of human property and status.[23] Yet mills also assumed a public character, and given the proper circumstances farmers and millers formed a symbiotic relationship upon which settlement prospered. Such facilities were among the first constructed on the frontier and were regarded as the vehicle by which settlers were uplifted from mere subsistence status to that of agricultural producers. For this reason inducements of free land were commonly offered to millers by the immediate township or a higher government level. Thomas Need recalled that his Bobcaygeon neighbours "worked together in great harmony and good will, notwithstanding their different stations in life.... [T]he completion of the [Need's] saw-mill was an event of vast interest to [those] ... who looked to exchange their rude shanties ... for neat frame houses."[24] The key to this equation was a strategy that carefully balanced the needs of the community with the aspirations of the capitalist, a symmetry that was breached in Ops Township.

While Ops pioneers were attempting a settlement in the Scugog lowlands, William Purdy was concluding his business affairs near York and planning a new venture. Purdy was the son of United Empire Loyalists who had moved in 1787 from Westchester, New York to Elizabethtown, Upper Canada. In 1814 he purchased a large mill complete with a ten-foot high and over 200- foot wide dam in Vaughan Township, north of York. He made several improvements to the structure, including the addition of a flour mill. After a fire destroyed the business in 1828, Purdy set his sights on rebuilding his mill. The source documents do not reveal why he chose to leave Vaughan or why he chose Ops. Perhaps in addition to being an entrepreneur, he also possessed a "pioneer spirit." By December 1829 the sixty-one-year-old Purdy petitioned for the lots that had been set aside by the province. Within four months he moved to Ops with his wife and adult sons, Jesse and Hazard.[25]

In February 1830 Purdy detailed his plans for the township: he wished to build a home and then construct a twelve- to fourteen-foot high dam across the Scugog watercourse. He would then attach a commercial sawmill and a common gristmill, both of which would be operational by October 1831. To enable boats to pass up and down the waterway he would create a slide in the dam. Purdy also requested that a reservation be made around these lots so that he could overflow adjacent lands while storing water.[26]

Peter Robinson was eager to locate Purdy in Ops, and he urged that the miller's offer be accepted by the government. He believed the dam would facilitate navigation by raising the water level, a valuable improvement for a proposed agricultural community with no secure means of transporting produce to and from Peterborough markets. Regarding Purdy's request for surplus lands, Robinson advocated that a

100-acre tract west of the proposed site be granted, at lot twenty-one in the fifth concession.[27] Purdy and his family moved onto the Ops lots in March 1830, and by 1 April had begun constructing the stone dam. A spring freshet swept the dam away, but by May the Purdys and several hired hands had resumed work and they would have rebuilt the structure were it not for a late-August visitation by the ague. Before being stricken himself, Purdy cleared at least five acres, put in a crop, quarried out a place for the sawmill and flume, and erected a wooden frame for the dam and mill. In light of the sickness that delayed the construction, Purdy unsuccessfully petitioned the government for relief.[28] Throughout the autumn of 1830 Purdy's dam neared completion and within the township a total of a hundred families had been located. During the next month the Ops community met and elected town officers. By 1833 nearly a thousand people resided in Ops.[29]

The spring 1831 freshets, however, were no kinder to Purdy than those of the previous year. An April flood swept away a portion of the dam, thus preventing him from operating the sawmill. The reserve of water that had been building up since December 1830 was also entirely lost. In writing to Lieutenant-Governor Sir John Colborne, Purdy conveyed this information as an explanation for the delays in the lot improvements. He expressed his own disappointment as well as that of the settlers and projected a July 1831 start for rebuilding.[30] A subsequent correspondence from Robinson to a superior reinforced his desire to retain Purdy's services in Ops despite the delays. Robinson hoped that the "most liberal consideration would be put upon [Purdy's] operations in fulfilling his agreement."[31]

Problems stemming from the dam became apparent to the miller's neighbours by late 1831. Purdy recognized that land adjacent to his mill dam would become flooded, which is precisely why he requested additional lots in his original application. In the process of storing water, several upriver lots became ruined. McDonell took note of the situation, informing Robinson that he was "aware that you will have several applicants ... some of whom will claim remuneration for the loss of a greater quantity of arable land than they have actually sustained, and that on examination the marsh being overflown, will be found the principle [*sic*] cause of complaint." Yet McDonell had little sympathy for the injured settlers, asserting that this was "a very small sacrifice on the part of the petitioners, when compared with the great advantages of having a mill at their doors."[32] It is unclear whether Robinson responded in any way to the petitioners. However, the grievance voiced by some settlers would serve as a flashpoint to the uneasy, often tragic, adjustment that Ops inhabitants would make to this situation.

Throughout this period Purdy's mill also prompted changes in the way people regarded the Scugog landscape. When Purdy arrived to inspect the proposed millsite and lots in 1829, the watercourse that ran through the township was merely a "stream."[33] Three years later it had grown. In 1832 the Purdys and twelve other petitioners wished to construct a road northwest from the mill, but their petition met with resistance from the remainder of the township freeholders. Those in opposition claimed that the road already under construction from the mill east toward Peterborough should continue, and that the alternate road proposed by Purdy and others would bisect many of their lots. Moreover, because the swamp so pervaded the area, any decrease in arable lands would further hamper their ability to support themselves. They also believed that Scugog "Creek" was the swamp's source.[34] This terminology is significant because townspeople gradually referred to the Scugog first as a meandering "stream," then a "creek," suggesting a steady and noticeable expansion in the watercourse's width. Finally, in 1834 and 1836 technical reports allude to the Scugog "River" when citing landmarks for road improvements.[35]

Second Nature:

The Trent Canal, Purdy's Mill, and Threats to Home Place

THE years 1833–36 were pivotal ones in the Trent Valley. During this time Purdy's relationship with Upper Canadian administrators soured as his mill conflicted with the progress of the Trent Canal. Simultaneously, local opinion increasingly turned against Purdy as more upriver inhabitants realized the repercussions that stemmed from the mill dam and pond.

In the 1820s the provincial government began funding large canal projects which would facilitate commercial navigation throughout Upper Canada.[36] By the early 1830s boosters in the back townships of the valley were advancing the idea of a canal system linking Lake Simcoe to Lake Ontario at the Bay of Quinte. The proposed route would pass through the Trent's watershed and more fully connect the district to the rest of the province. The plan for the canal was conceived in 1833 at Peterborough among prominent area men, all of whom were leading lights in their respective communities within the valley and who would also serve on the Commission for the Improvement of the Navigation of the Inland Waters of the

District of Newcastle, the political designation of the Trent Valley. The commissioners effectively wielded unlimited power to improve the area's resources, and no judicial oversight existed. The likelihood that conflicts of interest would colour the commissioners' decision making appeared not to have occurred to the provincial legislators, who earmarked $10,000 for the project.[37]

The commissioners approached their task not as disinterested community members but rather as land-owners, speculators, capitalists, and politicians eager to further their own interests while promoting the canal. Among them were Members of the House of Assembly George S. Boulton and Colonel Alexander McDonell. The commissioners wished to connect smaller watercourses, such as Sturgeon and Scugog Lakes, to the larger canal. This part of the plan attracted support from the back townships' farmers in Ops, Fenelon, and Verulam, who were especially isolated due to the lack of dependable navigation.[38]

However, a problem existed with the Scugog: Purdy's mill dam stood in the way, and he had no intention of sacrificing his concern for the good of the canal. On 16 November 1833 the commissioners petitioned the lieutenant-governor, arguing that "unless a lock [is] constructed at Purdy's mill ... a most valuable tract of land [Ops, Cartwright, and Manvers townships] will continue to be shut out from settlement."[39] One week later James Gray Bethune, a commissioner and local steamboat magnate, urged the legislature to grant additional funds so that heretofore obscured waterways at Cameron's Falls in Fenelon and on the Scugog and Pigeon Rivers might be opened.[40] Clearly, Purdy's mills complicated the designs of a large interest group, and from its earliest inception the commission did not look favourably upon Purdy's situation. Ops farmers also resented this impediment, which isolated them from Peterborough markets. In 1835 they petitioned for the establishment of annual fairs in Ops, citing as a key reason the "great inconvenience felt throughout the back townships in procuring stock of all kinds and particularly working cattle."[41]

By October 1833 Purdy's saw- and gristmills were operating. Having fulfilled his commitment to improve his lots and erect a milling facility, he petitioned Colborne for the patent deed to the property. Purdy also reiterated his desire that the deeds to the lots adjacent to his mill and millpond include a clause protecting him from damage claims. "[W]ithout such security," he explained to the lieutenant-governor, "[my property] may be much deteriorated and the mills [will] become valueless."[42] On the same day that Purdy petitioned Colborne, Alexander McDonell wrote to the Crown Lands Department to certify that "[the claim about] the mills and other improvements mentioned [in Purdy's] petition is correct."[43] It is significant that in his letter McDonell did not refer to any special consideration for Purdy's mill privilege, including a protecting clause.

Not until 6 May 1834 was Purdy's petition acted upon. Crown Lands Commissioner Robinson not only urged that Purdy be issued the deed for his improved lots, he also recommended that all future land deeds in Ops carry a reservation preventing damage claims against Purdy.[44] Three days later an executive order-in-council granted Purdy lots twenty and twenty-one in Ops' sixth concession; however, the grant made *no* mention of a protecting clause.[45] Thus, Purdy had received the deed to the land he had cleared and improved, yet failed to convince the administrators of his need for legal protection. The state of the law would have held Purdy liable for any damage claims brought by his neighbours; had he received the protecting clause, the claimants would have been powerless.[46] In Purdy's mind, the government had failed to grasp the magnitude of his efforts. His subsequent petitioning for unlimited control in directing watercourse matters only fuelled the discontent and divisiveness in Ops and the surrounding townships.

In erecting the dam, Purdy had raised the water level of the meandering Scugog. For many upriver inhabitants this had the disastrous effect of flooding their lands. However, one positive result was the improved navigation enjoyed by those Ops residents situated away from the watercourse, and by people in the neighbouring townships of Mariposa, Cartwright, and Reach, who benefited from the increased size of Lake Scugog. The fortunes of Port Perry merchants, for example, were buoyed by Purdy's dam because it raised the water level of the lake and facilitated commercial travel. However, the displaced farmers grew more discontented and contemplated legal action to remove the obstruction, a move countered by 254 Ops residents and settlers in the surrounding townships who joined with Purdy in petitioning that the millpond be maintained.[47] The petition was read and referred to the commissioners overseeing the improvement of the district's inland navigation on 22 January 1835. On 2 March, at the recommendation of the commissioners, a full investigation of the Ops matter was ordered.[48] During that same week another petition regarding the dam was read. Robert Jameson, the attorney general and a resident of Fenelon Township, north of Ops between Balsam and Sturgeon Lakes, petitioned the legislature to begin a survey of the waters between Scugog and Sturgeon Lakes. The purpose of the survey would be to make a plan and suggest lock sites on the fifty-mile stretch. Jameson intimated that the surveyor should also investigate the possibility of linking Ops with the Trent Canal, ascertain whether the dam and large millpond would hinder such a plan, and probe the damage claims

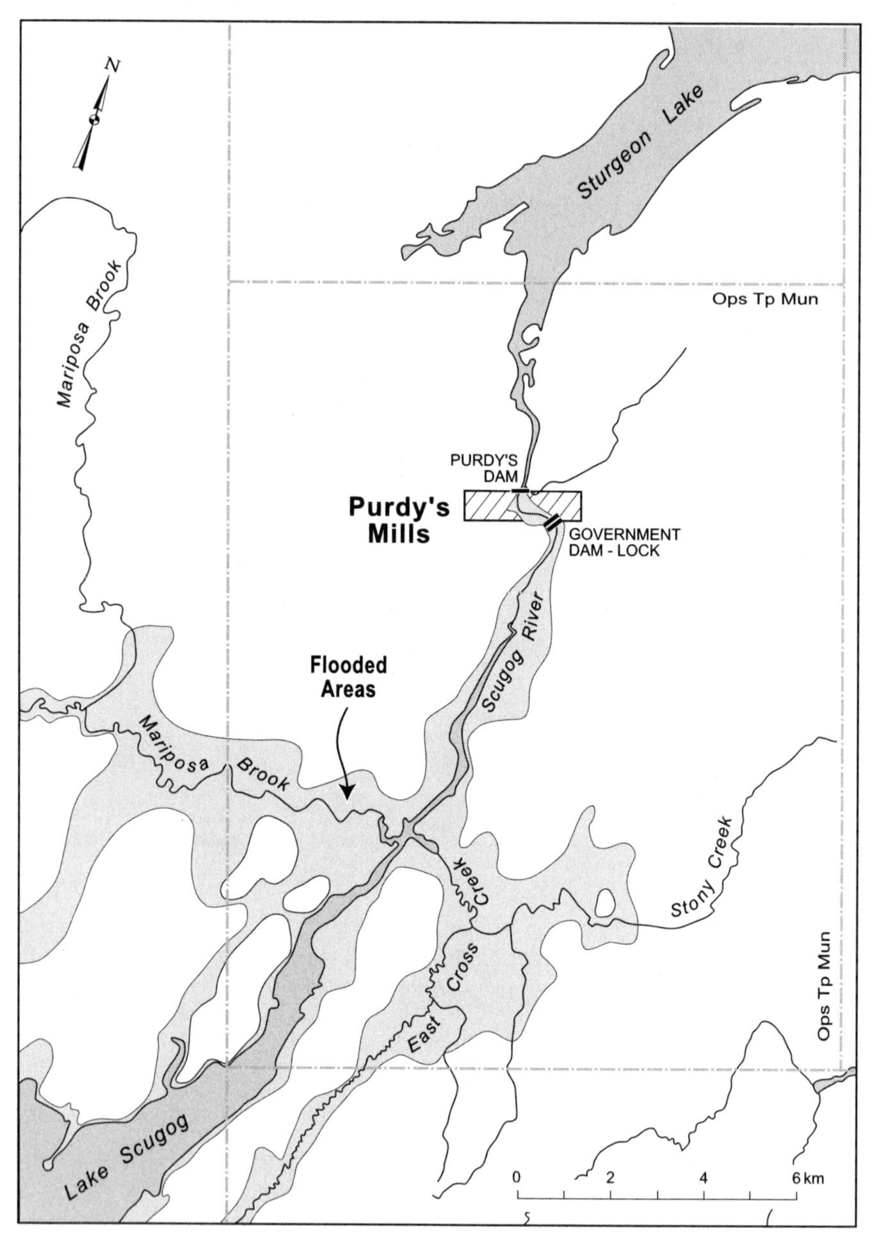

Figure 7: *Ops Township showing Purdy's Mills and flooded lands,* c. 1835.

against Purdy.[49] Similar sentiments were echoed by Assembly Speaker Marshall Bidwell in an address to the lieutenant-governor. Finally, a surveyor was dispatched to Ops.[50]

Nicol H. Baird was the Trent Canal's chief field surveyor and engineer.[51] He arrived in October 1835 and was shocked at the extent of damage. "As respects ... the extent of overflowing," he reported, "it is upon a much more extended and destructive scale than ... can be found from [a similar sized dam] in these Provinces."[52] He noted that prior to Purdy constructing his dam, navigation on the sluggish Scugog was limited. The Purdy dam had remedied this problem by raising the water level on the upper portion of the river, where previously only small craft could navigate. Unfortunately, the dam had also inundated small upriver waterways and drowned out other prospective mill sites. Specifically affected were the East and West (Mariposa Brook) Cross Creeks, which emptied near the mouth of Lake Scugog, as well as creeks in neighbouring Cartwright, Reach, Brock, and Mariposa Townships.[53]

The ruination encompassed an area thirty miles in length and width above the site. While travelling the nine miles from the mill site to its lake source, Baird observed that nine feet of water covered hay meadows and crops amidst a "continuous scene of drowned and decayed timber ... [which at intervals revealed] the former residences of the settlers, showing part of the roofs out of the water, from which the inmates had to make their escape." [54] Lands, crops, roads, and fishing sites were damaged, and desperate citizens now turned to canoes for transportation. The agricultural community had been obliterated by the very facility that was intended to aid the farmers. Purdy had also halted fishing on the river by placing traps at the mill and selling the captured fish at two pence each with "considerable annoyance to the public."[55]

Baird may have inadvertently crystallized public opinion against Purdy by elaborating on the topic of canoe transportation. He flatly stated that, as a result of constant canoe paddling, effluvia was being stirred in the water, which ultimately released bacteria into the air and caused the ague, or "lake fever."[56] Although it was found, in fact, that the mosquitoes carried and spread the malaria, Purdy's creation of more swampland and a permanent millpond certainly perpetuated, if not increased, the scale of the ague by providing the vectors with an ideal breeding ground. Baird's suggestion nonetheless contributed to the inhabitants' acrimony toward Purdy.

Baird also intimated that the miller was selfishly and inefficiently hoarding water resources to the obvious detriment of the community. Purdy had always maintained that his mill would not operate without an extensive head of water. As an engineer, Baird disagreed, asserting that "the grist mill is upon the rudest possible principle, constructed without any regard to economy of water, using as much and wasting more than would drive six manufacturing runs."[57] He maintained that with proper construction the mill should require only one-third of the water it presently used, thus alleviating the distress of those upriver.

Baird was less critical of Purdy when contemplating the reasons for locating the dam at its present site. Upon interviewing the miller and some original inhabitants, he learned that the location was considered to be the only one suitable for a mill privilege. This opinion stemmed from the fact that the Scugog turned to rapids as it crossed lots twenty and twenty-one of concession six. Baird concluded that although he was "disposed to give Mr. Purdy credit for having located the dam more to his own than the public advantage ... it would appear ... that there was no intention of overflowing so much or almost any land by the erection of the dam, it being the general opinion of the country around, who assisted Mr. Purdy in his arduous undertaking in a back county, that 'if he raised it as high as a house he would do no damage,' ... that Mr. Purdy, or anyone else, had the most distant idea of the ultimate result."[58]

However, a case could also be made against Purdy that in creating such a large millpond he had effectively undermined any competition to his mill. The Crown Lands Department chose the site for the mill in its earliest survey; however, Purdy began flooding adjacent lands and inundating the surrounding watercourses after he arrived, which prevented any future challenges to his concern. As well, a miller with Purdy's experience must have realized that the point at which the stream turned to rapids would be lost through the creation of such an extensive millpond. Purdy did anticipate such a reserve of water, which is precisely why he petitioned for additional land adjacent to his own lots.

After investigating the situation, Baird gave his recommendations for alleviating the distress. Since the complete removal of the dam would only cause the Scugog to lapse into its original, meandering, and unnavigable state, he advocated that the mill be maintained at its present location and the dam lowered by half (from fourteen to seven feet). The lowered dam would enable the river to revert back to rapids as it passed through Purdy's concern. As a corrective to the overflowing, Baird proposed placing another dam above the present one. This additional dam would be connected to a lock and would siphon-off the stored water which was flooding the upriver lots. The lock would also facilitate navigation along the Scugog. By lowering the overall water level by about one-half, farmers' land would be significantly drained and the spoiled lots reclaimed.[59] From the Baird report it appeared that all the lingering problems were remedied and that all parties were left satisfied: Port Perry merchants retained their transport trade, the canal commissioners got their Scugog connection, and the citizens' distress was calmed. All but Purdy were provided for, though the family would in time be given $1,600 for their inconvenience.[60]

Baird's recommendations were reasonable, yet they masked his true intentions for the Scugog. The engineer had his own agenda. Like the commissioners, Baird was searching for the best way to link the Scugog with the Trent Canal. His motivations were perhaps more personal and less profit-driven than those of the commissioners. Having invested the past several years surveying the watershed and drafting plans for its development, the engineer had a creative stake in its completion. As chief engineer and supervisor, Baird must also have realized that another work site along the canal would ensure a steady and perhaps extended salary for years to come. In 1836, for example, his pay for supervising the Trent River works alone was $2,500, and he earned another $1,250 for administering the other inland waters improvements.[61] It was in his best interest to advise that Purdy be brought into line with the wishes of the commission; given these circumstances, it is not surprising that he recommended lowering the dam, adding a second, and installing a lock to facilitate navigation. In Baird's opinion the situation at Purdy's Mills had ironically created "one of the most favourable opportunities ever presented, to open up the same extent of country by so little assistance of art, as the waters of the Scugog River and Lake afford." The Newcastle District's waters could continue to be improved, in this case to accommodate large-sized steamers, and more locally the townships immediately bordering Lake Scugog could benefit from access to the canal system. All of this could be realized "at a trifling outlay" of 2,500 pounds.[62]

Baird's recommendations appeared to mediate the conflict between the millowner and the aggrieved farmers, and also created a pragmatic basis upon which the wider canal project could be justified. More significantly, Baird's report brought to fruition the commissioner's desire to interrupt Purdy's plans by connecting his mill dam with the canal. In early December 1836 the Committee on the Newcastle District Internal Improvements adopted the engineer's recommendations.[63]

In Ops, magistrate John Logie had been circulating a petition against the miller, which he hoped would "shut old Purdy's mouth." He charged that the "old fellow wishes all the water to his own mill, and is against every improvement unless it serves himself."[64] The document was filed in December and its sixty-eight signers demanded that Purdy lower his dam by six or seven feet. In response to this challenge the intractable miller and 135 supporters petitioned the House of Assembly to have the canal bypass the Scugog completely and instead connect with the Trent River, thus preserving the status quo at Purdy's Mills.[65] Evidently, Purdy was not prepared to cooperate with his neighbours or allow the commissioners to overrun his mill privilege. His inflexibility would bring him much grief during the coming year.

Crafting a Defence of Public Rights

IN the years leading up to the controversy surrounding the dam, William Purdy fancied himself a beacon of the community, even though his neighbours thought otherwise. His milling operations were a key component of the new settlement, meeting the local demand for sawn lumber and milled grain.[66] By facilitating navigation Purdy had also cultivated a basis of support among the Scugog merchants, yet as events unfolded it became clear that this support was not genuine. However much Purdy saw himself as a "town father" he was hardly a "father of the people." In fact he had little in common with his neighbours. Being of loyalist parentage, a Methodist, and lately a capitalist just north of the province's most developed city, Purdy had no organic connections to Ops' Anglo-Celtic, agrarian-based pioneers. He was also a relative newcomer to the community.

Politically it is unclear where Purdy's allegiance lay: it is not known whether he was a government supporter or a reformer. In the Trent Valley it appears that Purdy did not have the blessing of his peers. He was never nominated for magistrate, commissioner of the peace, or health officer, all of which were locally high profile appointments and were accorded some measure of distinction.[67] In Ops he served as a commissioner on local road appropriations, as postmaster, and later as chair of the township council.[68] It was while he served in this latter capacity that public outcry against the miller turned to mass action in January 1837.

At the 2 January 1837 meeting of the town council, the bitterness toward Purdy was clear as freeholders gathered to select a new slate of township officers. The divisiveness created by the mill dam controversy no doubt coloured the meeting, and the mood soon turned ugly. Purdy chaired the session, though he acted contrary to custom. He would not agree to the acts advertised by the town clerk, he hurried the hearings, and he refused to read the town statutes even though several people requested that he do so. To further anger those assembled, Purdy abruptly terminated the session while the voting was in progress, denying "at least four-fifths of the inhabitants their just privileges of giving their votes."[69] He and the town clerk then adjourned to his own home to conclude the proceedings. It is possible that Purdy halted the voting because his own seat was clearly in jeopardy. In a petition to the district's justices of the peace, Logie led seventy-seven "true and Loyal Subjects" in formally protesting the undermining of "British laws and customs" by Purdy and twelve unnamed others. Logie punctuated the grievance by suggesting that "such conduct is likely to produce a feeling in the Township of a most mischievous tendency."[70] An accompanying petition included the names of the town officers that the Ops citizens subsequently chose; Purdy's was not among them.[71] By the year's end the "mischievous tendency" portended by Logie would become evident.

Following this incident, an aggravated Purdy pleaded his case to Sir Francis Bond Head, the lieutenant-governor. He underscored that at considerable cost, he had improved his lots and constructed mills that accommodated the public. Purdy also alluded to the arrangement that he and Robinson had discussed in 1834, whereby he was to be given permission to overflow adjacent lands "in order to have a convenient and advantageous use of the mill."[72] In fact, Purdy was *never* given such unlimited license: the arrangement that he and Robinson had discussed was not approved by the executive council. Purdy nonetheless urged Head to unconditionally reserve the adjacent lands for him and his milling activities as per the discussed terms. The lieutenant-governor did not act on the petition, instead referring the matter to McDonell, who confirmed that Purdy had "built mills in every respect adequate to the demand of the surrounding country."[73] However, McDonell knew of no understanding between Purdy and the government that would have allowed the miller to overflow other lands with indemnity.[74] Purdy received no satisfaction from the government, and plans progressed toward the building of the dam and lock above his mills.

During this time the now sixty-eight-year-old Purdy, weary from disputes with the government and his community, conveyed his mill property to his sons Jesse and Hazard. In November the elder Purdy was urged to name an arbitrator and meet with the government's representative to negotiate a monetary settlement for his loss of waterpower resulting from the lock's construction. The commissioners also planned to attend the Ops 18–23 December meeting.[75] It never took place, however, due to circumstances that neither Purdy nor the others could ever have imagined. Aggrieved Ops citizens had yet to receive relief from their dire situation, and their rancour now reached its apex. Soon their wrath would coalesce with another movement astir in the province.

The rebellion in and around Toronto (formerly York) prompted the mobilization of loyal militia throughout the province. The Trent Valley was a stronghold of support for the conservative government during the turbulent 1830s.[76] Colonel McDonell commanded the Second Northumberland militia; on 5 December 1837 he was ordered by Bond Head to summon his officers and troops and proceed to the capital.[77] In the following days men from throughout the district mustered at Peterborough. One of these men was Major Thomas Murphy, who had recently purchased an acre near Purdy's mill site and had opened a small store there. In addition to his ambitions as a merchant, Murphy also sought to unseat Purdy as the township's postmaster.[78]

Ironically, it was Murphy's personal motives that served as the catalyst for Purdy's removal from Ops. While in Peterborough, Murphy began an unfounded rumour that William Lyon Mackenzie, leader of the Upper Canadian Rebellion, was hiding at Purdy's Mills. The local press embellished this hearsay and reported that the rebel leader had retreated to the countryside "to rob a mill."[79] In reality Mackenzie was either in, or en route to, New York State.

On 12 December, Colonel McDonell and his troops left Peterborough for Port Hope, where they were to be transported by steamship to Toronto. However, once the militia reached Port Hope, McDonell received a despatch countermanding the order.[80] Perhaps acting on Murphy's story, the militia then marched to Purdy's Mills in search of rebels. At nightfall the small army of 300 men descended upon the town with their flag unfurled, startling farmers and their oxen at the mill and awakening the townspeople with musket fire. Cheers were raised, trumpets sounded, kettle-drums rattled, and more guns were fired; however, no rebels were to be found. Tired from the day's trek, the group pitched camp around Jeremiah Britton's tavern.

The next day Major Murphy made his move and charged William Purdy as a reform conspirator.[81] Clearly the charge had no validity. It is suspected that during Baird's 1835 investigation, Purdy voiced his displeasure with the government for not allowing him to flood adjacent lots. Purdy may have spoken ill of the Family Compact, but his affiliation with reform elements was circumstantial, if not pure speculation. Nevertheless, on 13 December Purdy was arrested for treason and imprisoned at Cobourg until 3 January 1838, when he was discharged without trial. The local press further vilified the miller when it reported that "several persons have been committed to our prison ... among the number, Mr. Purdy ... who had received great benefit from the government."[82] Murphy achieved his goal and succeeded Purdy as postmaster, an instance of petty backbiting orchestrated amidst a carnivalesque atmosphere of confusion.

The charge of treason is as significant as it is curious. Though the accusation had no basis in fact, it appears that no one came to Purdy's defence. Neither Colonel McDonell nor magistrate Logie objected to the charge. Given McDonell's position as a member of the House of Assembly, it is only slightly possible that he did not want to interfere in what could be considered a township matter. However, it no doubt occurred to the colonel that if Purdy were removed from Ops, or at least coerced into changing his position, the situation might benefit McDonell and the other commissioners. Logie and others already doubted Purdy's respect for British custom, law, and institutions and there was no denying that they objected to his mill dam and the damage it caused. Thus, in allowing Purdy to be arrested, Ops people may well have been protecting their concept of "public rights" and "community," something the state seemed unable or unwilling to do. They easily borrowed a label from contemporary discourse that would cap Purdy's character assassination. In Ops, "reformer" and "traitor" appear to have been synonymous terms. Purdy's self-image of Ops town father was also shattered. He had built a base of support among those unaffected by the flooding, but when dissent turned to action, the miller learned that this bulwark was as unstable as the swamp that mired the Scugog lowlands. Perhaps the "Purdyites" also feared reprisal: they were nowhere to be found.

Ops people allowed the charge against Purdy to go unanswered, and in so doing they settled an old score with the miller. Purdy's arrogant behaviour, flooding and devaluing of lots, constant petitioning for further privileges, opposition to the proposed canal, and misconduct at the January 1837 town meeting had galvanized public opinion against him and was clearly an integral precursor to the community's silent verdict during the rebellion. Purdy continually eschewed public opinion and ushered in the township's mood of "mischievous tendency" alluded to by Logie. Ops citizens sought an appropriate moment to act against Purdy, and the 1837–38 rebellion provided a fulcrum for change.

Following his 1837 imprisonment, William Purdy moved from Ops to Bath, near Kingston, where he spent his remaining years as a farmer and died in 1847 at age seventy-seven.[83] His sons Jesse and Hazard remained in Ops and unsuccessfully tried to sell the mill in the spring of 1838.[84] Soon after, Jesse also moved from Ops, leaving Hazard as the sole proprietor. By January 1839 Hazard Purdy had not acted on any of Baird's recommendations: the dam stood at fourteen feet and the millpond remained. The commissioners contemplated legal measures against the miller, but these were never acted upon.[85]

Nature Responds

THE construction of the government dam and lock above the Purdy mill dam began in 1838, though the area swamps and forest impeded progress.[86] The flooded forests produced poor quality timber, which, for want of better, was used in construction at the Scugog and nearby Bobcaygeon sites.[87] The huge millpond also complicated construction, and by autumn 1839 workmen were continuously hand-pumping backwater from the site.[88] A shortage of labourers only added to the problems. Area men were recruited as navvies, but many left their jobs to tend to their fall harvests. In fact, many of these farmers sold potatoes and hay at the work site.[89] A more obvious explanation for the delay stems from the paucity of funds available to the contractor, Homer Hecox, who eventually abandoned the works and returned to the United States, taking with him some of the navvies' wages.[90] As a result of chronic delays in payments, workers were incited to petition Baird for their wages in June 1840.[91] In spring 1841 construction remained at a standstill.

It is probable that timely completion of the dam and lock, and the consequent reclamation of flooded lands, would have reduced the stagnant water. However, this did not happen and the water was allowed to remain on the flooded lands into the summer of 1841. The 1841 spring floods had been particularly heavy in the Trent Valley. Despite the force of three separate freshets it was proudly reported that none of the government-built dams, including that on the Scugog, had "failed in the slightest degree" to hold back the excess water.[92] This was welcome news to the Trent Canal boosters, but not for Ops citizens.

The wet spring was subsequently followed by an extremely hot and dry summer, which threatened area grain and hay crops. The excessively warm temperatures continued into the July harvesting period. In late August the local press reported that as the heat wave continued, sicknesses had occurred in the back townships.[93] The press was unaware of the scale of distress in Ops; indeed, the remoteness of the rear township helps explain why no further news of the sickness was forthcoming. The prolonged flooding, followed by drought and excessively warm temperatures, provided ideal breeding conditions for the mosquitoes. The insects fed on the blood of the settlers, infecting them with malaria on an unprecedented scale. The fever lasted from July until December 1841.

The Canada Company followed the Crown Land Department in locating persons in various townships throughout Upper Canada, and by 1841 owned many of the land patents in Ops. The company was understandably alarmed by the devaluing effect of the flooding in Ops and area lands. Frederick Widder, the company's commissioner, later conveyed this sentiment along with the details of that fateful summer, to J. B. Harrison, the governor general's secretary. Widder laid the blame for the ruined lots and the unhealthiness of Ops squarely at Purdy's door, or rather, mill dam. "In one place," he reported, "the heads of forty families, besides women and children died."[94]

So dire was the situation that there was no one left to bury the dead, and as many as six days passed before anyone saw to removing the bodies. Some children and untended ill people starved to death after their caregivers had died. In the most extreme case an entire family was discovered dead in their home, and such a putrid stench permeated the area "that the neighbours found it necessary to burn the house, corpses and all, as no person could be found to enter it for the purpose of interning [*sic*] the bodies."[95]

The ague also prevented Scugog inhabitants from paying their land patents. The Canada Company received applications for overdue instalments, and some settlers were only able to forward a few dollars in payment, citing the death or illness of family members and their own inability to harvest crops. "Others," Widder reported, "have come to the office and personally made the same excuses, and their unhealthy and emaciated appearance has been a sufficient warrant for the truth of their statements."[96]

From his sources, Widder confirmed Baird's conclusion that Purdy had constructed his dam in a most advantageous location. According to old settlers, "Purdy erected his dam where it is, instead of a few hundred yards further down the river for the purpose of destroying the only other mill sites in the neighbourhood, and in which he has perfectly succeeded, as six powerful mill sites have been rendered useless, within a few miles of [Purdy's Mills]."[97] In reality the provincial government chose the mill site, though Purdy chose to construct his dam at such a height as to create an extensive millpond, thus overflowing potential upriver competitors.

Following this period of extreme distress, two separate riots ensued. The first occurred in early December when a crowd believing that Hazard Purdy's mill was responsible for the deadly fever attacked and destroyed part of the dam. Following the incident, Logie related the rioters' complaints to the governor general. The matter was eventually dismissed, the administrators believing that the sickness was merely the latest wave of ague endemic to the Scugog area. Enraged by the lack of serious interest in their case and still suspicious of Purdy's mill, the crowd again assembled on 27 December. Their ranks swelled as farmers from neighbouring Cartwright and Manvers Townships joined the angry mob, bringing with them rifles, pitchforks, and axes. Despite Logie cautioning the crowd that further violence would only harm their cause, the armed mob "left ... and set to the march of destruction and completely destroyed the saw mill flume and remainder of the dam."[98]

The mob was only partially correct in targeting Purdy's mill dam as the main source of the sickness. Ague had been an almost regular occurrence each August in Ops, even before William Purdy arrived. However, the ecological damage caused by widening the Scugog watercourse, enlarging the lake, and fostering mosquito breeding grounds certainly worsened an already unhealthy situation. Nonetheless, as Widder asserted in the wake of the riot, those who were so "seriously injured by the dam ... [would] proceed to any lengths, rather than permit the reerection of it. They are too poor to proceed to protect themselves, by those means which the law affords, the ultimate course must be evident."[99] Ops citizens had waited patiently for a lawful solution to their problem. When no genuine action to alleviate their suffering appeared, they moved from a peaceable political vehicle like petitioning to a direct and more traditional form of public action in defence of their rights.

In the aftermath of the riot and dam-breaking, a committee was formed to investigate the situation at Purdy's Mills. The committee concurred with the rioters' questionable opinion that flooding by the dam had caused the 1841 summer sickness. It recommended that if the dam was rebuilt, it should stand no more than eight feet, noting that the mill could be sufficiently supplied and that navigation could also continue. More importantly, a lowered dam would allow for land reclamation. The committee also recommended that floodgates be installed so as to allow excess water to be occasionally drained.[100] Hazard Purdy would be allowed to reconstruct a dam of lesser height, which would allow the Scugog inhabitants to fully develop their lands. Committee chair John Langton summarized that "such a measure would give the greatest satisfaction to the whole population, who have so long suffered from the inundation; but [the committee] would consider it unjust to Mr. Purdy if remuneration were not provided for the damage and interruption of his present works."[101]

Langton felt compelled to address the riotous behaviour. He attributed the outburst to the "distress arising from sickness and not from any malicious feeling towards Mr. [Hazard] Purdy, of whose moderation and forbearance throughout all parties speak in the highest terms."[102] The provincial government agreed with the committee on the case of rioters, and on 1 March 1842 the provincial secretary instructed the justice of the peace to stay the fifteen arrest warrants issued against the principal, though unnamed rioters. The initial request came directly from Hazard Purdy under the condition that no further violence be taken against his mill.[103]

It appears that Hazard Purdy chose not to act on the committee's recommendations. By December 1843, with the government's dam and lock construction coming to a close, Purdy was granted 400 pounds, plus the use of all the surplus water not needed for navigation. In exchange the miller was to relinquish his damage claims and keep the new dam in repair. The government dam and lock were completed in the summer 1844, and shortly after, Purdy sold his business.[104]

Conclusion

WHAT occurred in Ops between 1826 and 1841 was a public struggle to define the township environment and a place of residence.

The resulting divisions within the community prompted citizens on either side of the mill issue to engage in various forms of "popular politics" to effect change. In identifying press activities, extraparliamentary organization, parades, rioting, and criminal activity as manifestations of popular politics, Carol Wilton underscores the omnibus nature of the term and explains how Upper Canadians effectively marshalled some or all of these features to their advantage regardless of class, ethnic, or political difference. Popular expression through petitioning, which many, including Ops inhabitants, gravitated toward, was also part of this larger constellation .[105]

The expression of popular politics seen in this case was unique to early Canada. Pre-Confederation violence usually involved political or ethnic groups or workers.[106] Clashes between farmers and millowners appear to be unusual. At Purdy's Mills historical antecedents played a key role. The citizens' attachment to English legal conceptions of due process helps to explain why the citizens did not attack the dam in 1835 when the flooding problem first arose. They instead petitioned to have the dam lowered. Ops people did not destroy the mill dam, the symbol of economic development, but rather lawfully challenged Purdy with petitions and finally used smear tactics in an attempt to oust him, or at least compel him to change his position. In this sense Ops citizens' actions appear more as a "charivari," communal actions taken to undermine social authority or show disapproval for certain forms of behaviour, than as an angry response to progress.[107] This explains why Ops citizens did not simply tear down the dam. They stood by as Purdy was routed from Ops through rumour and innuendo. Ultimately the militia, an organ of the state and commanded by an interested party (Colonel Alexander McDonell), was manipulated to achieve this goal.

The 1841 dam-breaking also requires elucidation. In the midst of much sickness and death, Ops and neighbouring township citizens attacked and destroyed the mill dam. Their actions do not reflect a popular revulsion toward development; indeed, they wished to become more integrated into the province's commercial economy. Citizens had recognized early that the dam facilitated navigation on the Scugog. This was a necessary first step toward linking the rear township farmers with the Trent Canal system. Ops people had a stake in the completion of this network, but Purdy's fourteen-foot high mill dam and the consequent overflowing and devaluing of their lands confounded such a plan. Baird elaborated on this position in 1842. Reflecting on his 1835 Ops inquiry, he noted that "seven years ago I called a meeting of the inhabitants who threatened to tear down the mill dam and would have done so but for the belief from the survey going on and a statement made that the water would be lowered."[108] Nicol H. Baird resurveyed Ops in 1842 and found a similarly angry and resentful mood among the citizens. However, he asserted that they again wanted the water lowered through a government dam and lock and were willing to endure short-term illnesses so long as the works were completed.[109]

The impetus for attacking the dam stemmed instead from the citizens' belief that the mill concern was the sole source of mass sickness. The Purdys' refusal to reduce the dam's height and allow land reclamation, and the lack of serious attention paid by the state were the major factors fuelling the rioters' rage. There were political disturbances in Mariposa Township and along the lower Trent River between 1838 and 1840 in the aftermath of the rebellion, especially in wake of Lord Durham's *Report*; however, Ops citizens did not participate in them.[110] In fact one contemporary reports that the township contained a majority 150 persons "loyal" to the government, while in neighbouring Mariposa only five to ten persons fit that profile and upwards of 250 people were considered "disloyal."[111] Such ideological skirmishes did not play a role in the 1841 dam breaking. Instead, citizens resorted to very direct means to solve a very specific problem. What is as significant in 1841 as it was in 1837 is that the province never brought charges against the perpetrators of these crimes, thus in effect condoning their behaviour.

Purdy's mill dam had forever altered the Scugog watercourse, and in turn the Ops Township landscape. The dam turned the slow-moving water of Scugog Creek into a river and greatly enlarged what is today Lake Scugog. A descendant of William Purdy recounted to a local historian how Purdy found the area in 1830. "Scugog then," he recalls, "no more deserved the name of lake than the shelter of night deserved the name of house. It was a marsh and grass, the only clear water being that in the channel followed by the scow."[112] In the late 1840s and through the 1850s, many of the drowned lands remained underwater, prompting some Ops families to successfully petition the province for compensation. In 1854 they left after being granted lands in Euphrasia Township.[113]

The case of Ops Township demonstrates the place of the environment in the shaping of human history. In no small way, nature—the swamps, mosquitoes, and pathogens—was a force in the unfolding drama. The various ecosystems within the Trent Valley were either receptive or hostile to human manipulation. The events in Ops also foreshadow the later government involvement in seeking to create communities in inappropriate ecological locales: the expenditure for colonization roads north of Bobcaygeon also involved a social matrix in which the state, settlers, and private interest worked to reshape the Trent Valley, with equally mixed results.

5

The Road from Bobcaygeon: Lumber & Colonization, 1850s–1870s

As we saw in the case of Ops Township, the environment of the Trent Valley could circumscribe more than encourage settlement. This same first nature versus second nature dichotomy of settlers attempting to impose their will upon the land was evident in a more northern locale, amidst the rich pine forests beyond the Kawarthas and along the rocky Canadian Shield.

In the mid-1850s Upper Canadian administrators pursued another ambitious plan, this time to colonize the lands skirting the Canadian Shield. They began underwriting the construction of roads to facilitate lumbering and settlement. The idea for colonization roads grew out of the Victorian belief that human energy and intellect could overcome many obstacles. Indeed, the key themes that captured the Victorian imagination were science, geographic expansion, and nationalism.[1] Bipartisan politics in Upper Canada (or more properly Canada West after 1840) embraced the road-building project, believing that such an ambitious land policy could expand agricultural settlement onto the southern fringe of the Canadian Shield.

Administrators looked to lumbermen to blaze new trails for permanent agricultural settlement. It was generally acknowledged that the lumbermen would harvest the forest's bounty and then move on, but the farmers attracted by the lumbermen and their camps would remain rooted in the land. Lumbermen provided such farm families with their first commercial outlets, as surplus crops were bought by the lumber camps. Viewed in this way, the colonization scheme was a form of social policy, one designed to provide the growing population of Canada West with cleared land.[2]

This had been the experience in the upper Trent Valley. Unfortunately, when the lumberers moved to more verdant forests, so too did the farmer's market, as no permanent entrepôts could be found. Peterborough and Lindsay markets lay too far south to make the trek profitable. Moreover, the region's thin soils yielded no specialized crops: the wheat, potatoes, corn, and turnips raised there were also grown aplenty in many other parts of the province. Though it was hoped that Georgian Bay and Parry Sound consumers could be more easily accessed via the government's colonization road project, this never came to fruition. As we shall see, the quest for adaptation on these lands would prove difficult if not impossible, since the Shield, like the Ops swamps, shaped settlers' decisions.

The Natural Setting

The Canadian Shield is one of the most marked geographic features on the North American continent. Forged during the Precambrian age more than three billion years ago, it is a place more suited to forest than to agricultural crops. In the mid-nineteenth century these lands were coveted by lumbermen and shunned by would-be farmers—that is, until population pressures in the province tempted some to attempt the colonization of these forbidding lands.

Straddling the Canadian Shield and situated on the rocky Dummer moraines physiographic zone, the Crown's "waste lands" at the north of Peterborough and Victoria Counties appeared to mark the reasonable limits of agriculture.[3] Indeed, the upper portion of the Trent Valley is characterized most by sand, gravel deposits, and eskers.[4] Yet Crown Lands Department surveys consistently provided conflicting information about the land's agricultural potential. Alexander Shirreff, for example, envisioned millions of quality acres available for farmers. In 1835, however, a government team concluded that Shirreff "had drawn too favourable an inference from the level and quality of timber, [while] neglecting the more important consideration of the soil which almost everywhere throughout the country appears to be excessively light and sandy and often very shallow."[5]

Second Nature: Regional Expansion and the Lumberman's Vision

COLONIZATION roads grew out of the 1853 Land Act, which authorized the government to make free grants of up to 100 acres. Invariably these state-financed roads, intended to facilitate wagon-travel, were constructed through unoccupied Crown Lands. Lots were allocated on both sides of the north-bound roads. The first comprehensive plan was broached in 1849 by Bytown's Crown Land timber agent, James H. Burke, ostensibly to aid the settlement of the Ottawa–Huron region. In the short term, reasoned Burke, the lumber camps would purchase the settlers' surplus crops. It was hoped that after the lumbermen had moved on, those same farmers could find markets in the Ottawa Valley or at the Bay of Quinte. Ultimately, however, this would not be the case.[6]

Lumbermen certainly benefited from this plan. Historians generally agree that dependable winter roads, built at government expense, aided the loggers' excursions into the forest frontier.[7] In the Trent Valley, Mossom Boyd championed the building of the Bobcaygeon Road in the late 1850s, seeing the project as a means to undercut his Ottawa Valley competitors. Boyd's mills at Bobcaygeon were better situated to receive the logs, which could then be cut, put onto lakes and rivers, and driven south in the spring. The Gilmour, Harris, and Bronson companies were all active in the area but did not enjoy this geographical advantage.

The colonization road projects coincided with provincial immigration initiatives. Crown Lands administrators and field agents worked toward the goal of locating emigrants and Canadians on remote tracts of land. One reason for the government's concern was financial. The projected sale of additional Crown Lands would generate revenue that could be used to improve canal and railroad construction or to lessen tax burdens. Furthermore, mass immigration to Canada was being siphoned off by attractive American land policies. In the provincial context, Canada West's residents, bordered by the burgeoning Midwestern American states, hoped to increase their population and offset cross-boundary deficiencies, as well as surpass Canada East through political representation.[8]

Colonization roads appeared to allay some of the anxiety with which administrators approached the out-migration problem. In order to rally interest among the masses Peter M. Vankoughnet, Minister of Agriculture in 1856, began publicizing the venture: "TO EMIGRANTS AND OTHERS SEEKING LANDS FOR SETTLEMENT. THE PROVINCIAL GOVERNMENT have [*sic*] recently opened out THREE GREAT LINES OF ROAD."[9] Vankoughnet detailed information that would assist interested persons in settling, including the names of granting agents and the conditions: settlers eighteen years and older had to take possession of their allotments within one month, cultivate at least twelve acres of land within four years, and build a home and reside there until their duties were performed. After these requirements were met, the settler would be granted the property's title. Although the advertisement mentioned only three roads—the Ottawa and Opeongo, the Addington, and the Hastings—seventeen others were also constructed between 1856 and 1866.[10] Among them was the Bobcaygeon, opened in 1856, which would eventually extend north from its namesake into present-day Haliburton County.

Within the Trent Valley, as early as 1851, opinion was divided as to whether or not the wastelands were habitable. Some in Peterborough, such as Sherriff Wilson Conger, found merit in the plan, while settlers more familiar with the back country expressed doubt. For example, one man from Galway Township, situated on the moraines fronting the Shield felt that the plan to "settle the siberian waste lands in the rear of this County is all nonsense."[11] The agricultural capacity of these lands was at best marginal. As would later be discovered, any permanent settlement had to be wedded to seasonal income such as that provided by road construction or forest work. Nonetheless, a clamour arose in Peterborough during the 1850s for the government to open such remote lands. Often the vehicle of expression was the Reform press organ, the *Peterborough Review*.[12]

Under these unique circumstances, the lumbermen were receptive to the road project. Traditionally settlers were the bane of lumbermen, occupying valuable forest tracts, limiting the culler's mobility, and threatening destruction through careless fire use. In this instance, however, Mossom Boyd of Bobcaygeon, the Trent watershed's most successful lumberman, felt otherwise. Boyd's extensive operations stretched from Emily Township in the south into the as yet unsurveyed lands beyond the rear townships. By the 1850s he held the license to cut timber on some unnamed lots and his cullers were taking pine and hardwoods from the forests north of Bobcaygeon. If plans to build a road from his village into present-day Minden in Haliburton County came to fruition, Boyd stood to gain easier access to such bounty.

Although Mossom Boyd is an important figure in this story, any discussion of the upper Trent Valley must begin with Thomas Need. An English gentleman and

Oxford graduate, Need was one of the first Trent Valley citizens to move so far north of Peterborough. In 1832 Need purchased 3,000 acres in Verulam Township, encompassing what is today Bobcaygeon.[13] Need recognized the opportunities of forest frontier settlement and in July 1834 erected the area's first sawmill on his land.[14] From his mill pine was sawn and sent to England.

The forest's density impressed Need. He marvelled at the stands of pine, beech, birch, maple, and basswood enveloping Sturgeon Lake. "The whole continent of North America," he recorded in his diary, "is covered with trees of various kinds of gigantic growth, and if we except the savannas of the west, the prairies of the north, and a few insignificant beaver meadows every spot where a tree could grow nurtures one of these mighty sons of the soil."[15]

The valley's economy was intimately rooted in the lumber industry. By 1851 Peterborough County contained twenty-five water-powered sawmills, which annually cut 11,589,000 feet of lumber. Victoria County's six mills produced a more modest 1,056,400 feet per year.[16] To a visitor, "it seems as if the desire were to shut out the free wilderness by every means.... War to extermination against the forest is the settler's rule; and thus the instances are very rare of groves of the primeval woods amid the rising town."[17]

Sawmilling continued into the next decade. In Peterborough County alone thirty-seven mills employing 637 workers cut 68,821,000 feet in 1860.[18] However, due to rising transportation costs, Peterborough was slowly losing prominence as the watershed's sawmilling centre. Perhaps as a local motivation for colonizing the northern reaches, it was asserted in 1867 that "[a]ll our older townships have now for some years been denuded of the valuable timber.... [N]ow, the new and comparatively remote townships in the 'back country', as it is called, are mainly to be depended upon for a supply."[19]

In the 1860s the town of Bobcaygeon stood on the fringe of the "back country." Thomas Need's contributions notwithstanding, Bobcaygeon's most prominent citizen was undoubtedly Mossom Boyd. His father was a Northern Ireland officer in the Bengal Army in India when Mossom was born there in 1829. Mossom came to Bobcaygeon in 1834, bought Need's mill, and between 1839 and 1845 rafted timber between the upper watershed and Québec.[20] In 1849 Boyd enhanced his operations when he entered into a short-lived partnership with John Langton of Fenelon Falls.[21] By 1858 Boyd's operation was cutting 20,000 feet of lumber per day, and his landholdings in and around the northern watershed continued to grow.[22] The business expanded when, in 1871, Mossom Martin Boyd joined his father in creating

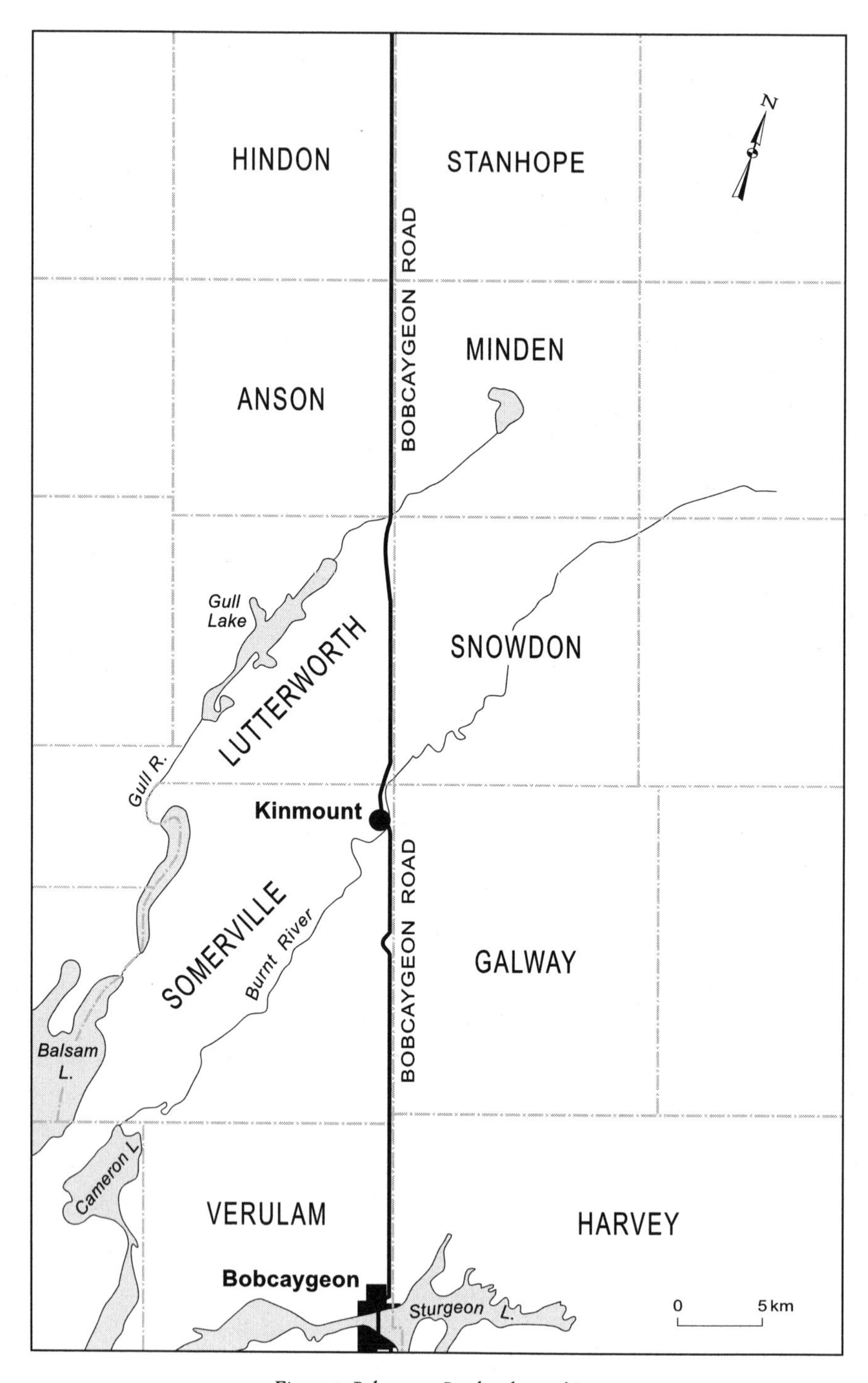

Figure 8: *Bobcaygeon Road and townships.*

one of the most integrated lumbering operations in central Ontario. The company continued to thrive even after the elder Mossom's death in 1883.

Second Nature: The Settlers' Reality

Occupational Pluralism on the Northern Frontier

IN the case of the Bobcaygeon Road, settlement preceded the official inducements to locate. In 1856, for example, the provincial Bureau of Agriculture advertised free land grants to "emigrants and others" on only three recently opened roads: the Ottawa and Opeongo, the Addington, and the Hastings. Although the Bobcaygeon Road had only five completed miles in 1856 and was nearly impassable beyond that, a steady stream of Canadians and newly arrived immigrants ventured north beginning in 1857.[23] By 1859 over 300 had arrived and located on free government grants in Galway, Somerville, Snowdon, Lutterworth, Minden, Anson, and Stanhope Townships. Adjacent to Stanhope, the Township of Hindon was formed in the early 1860s.[24] Under Crown Lands Department auspices and the superintendency of its agent, Richard Hughes, a "colony" took root on the northern frontier. The proposed road only hastened settlement.[25] Between 1858 and 1866 the Bobcaygeon project received $35,150.[26]

Sentiments expressed by area field agents seem to validate the desire to populate colonization roads with new immigrants. By October 1861 the Bobcaygeon Road extended about fifty miles north. Along the initial fifty miles the land was poor for agriculture; however, beyond that, in Snowdon and Minden, comparatively better soils supported oats, spring wheat, and potatoes. Crown Lands agent Robert McNaughton frequently travelled between Canada and Northern Ireland and successfully located some Ulster men in the back townships. On trips to Northern Ireland he displayed produce samples and urged others to join the new settlers. In addition, the settlers themselves wrote their former neighbours and related the positive experience of Canadian immigration. McNaughton believed that this type of advertising was imperative if Canada was to attract the best quality of emigrant. As he explained to the Commissioner of Crown Lands, P. M. Vankoughnet, "the class of Emigrants which we want in Canada at the present is farmers with a little means to enable them to proceed at once to *improve* the *land* and get themselves *comfortable homes*."[27] Moreover, McNaughton implored, "[P]ut me in a position to go back to Ireland [and appoint me as] an agent for some one of three Kingdoms ... [so as to continue to draw] a large industrious and (in some cases) rich emigration and not only to our shores but as settlers on our wild lands."[28] Indeed, suggested the agent, a forceful championing of Canada was required so as to counter the draw to the United States. In fact, Canada needed to follow the example of Australia and New Zealand, which "have their agents in every part of the old country handing out great inducements to emigrants."[29]

McNaughton went further by suggesting that the province enlarge its colonization offer and temporarily house new immigrants. Sheds to accommodate 100 persons, equipped with stoves, could be erected along the colonization roads and thus provide a base from which the settlers could build their shanties. Local woodsmen, earning one dollar per day could also meet the settlers and assist them in locating. Such assistance, reasoned McNaughton, would help stem the tide of discouraged persons who attempted to settle remote areas but ultimately exhausted their funds and moved back to established communities or went west.[30]

Richard Hughes, the agent in charge of the Bobcaygeon Road, identified similar problems. He lauded the short-term housing program, as this would keep the immigrants from having to spend money at local hotels or leaving for the western states. Hughes also concurred that McNaughton's Northern Ireland locatees were good settlers. He hoped that more families of this "superior class" would choose Canada; most had settled in the back country while others had first gone to earn money. Overall, Hughes reported, all free grants along the road had been claimed, and he hoped that more lots would be opened.[31]

Throughout the province, colonization roads also aided the lumbermen by providing them with passable winter roads.[32] It is perhaps not surprising, then, that Mossom Boyd accompanied Hughes on his treks through the north woods during the initial year of settlement. Both men travelled to Burnt River in northern Galway Township to inspect the road work being done and assess the requirements for a bridge. The Kinmount bridge, noted Hughes, would have to clear the water with enough room to allow timber to pass down.[33] Boyd soon followed up Hughes' report with a letter lauding Galway's pine and hardwood forest. Though satisfied with the prospects for settlement, he lamented the spectre of land speculation, which was keeping Somerville Township lands closed.[34]

The first families to move onto the lots encountered much difficulty during their initial five years. Reports filed by Hughes were mostly about the speed of land clearance. Each family or single man was given a free grant of 100 acres. During 1858 only four of the 168 family units were able to clear ten or more acres; the next year only twenty-four of 200 family units achieved that level.[35]

The settlers' resourcefulness, however, was evident from the outset. At the end of 1859, for example, Hughes reported that a total of 697 persons had been located and that families were engaged in a number of subsistence and commercial activities. In the 1859 season, wheat (1,620 bushels), potatoes (20,700 bushels), turnips (15,400 bushels), and corn (500 bushels) were grown throughout the seven townships. Settlers also cultivated beaver meadow and timothy hay; produced maple sugar and molasses; sawed lumber and shingles at Kinmount; barrelled potash; and traded venison and furs. The total value of all produce in 1859 was $20,790. During 1860 farmers added peas, oats, beef, and pork to their array of produce, thus increasing its value to $27,591.[36]

Despite the initially favourable farming pursuits, settlers needed to engage in other activities for continued survival. Although the land was free, ready cash was needed to buy livestock, farm implements, and other staples; settlers probably subsisted on more than simply potatoes and turnips, but the precise dynamics of the agricultural economy are unclear. Produce was consumed locally and in some cases surplus was sold to lumber camps. During the 1860s some settlers also turned to forest work and road labour as sources of income.

When the Bobcaygeon Road project and its accompanying settlement were proposed in the mid-1850s, Mossom Boyd held timber licenses on much of the valuable land along its route. His ultimate goal was to establish logging camps at Burnt River and Gull River, two northern sites that could be best accessed by the road. The Gilmours and others had been cutting timber on lands adjacent to Boyd's up to this point. They may even have been pilfering timber from Boyd on lands that were difficult to oversee. By including the road project and its intended settlement among some of his holdings, Boyd could better monitor his licensed berths. Moreover, the road itself held the promise of providing better access to Boyd's northern holdings. Thus his competitors were prompted to form a new strategy, one that would weave the free grant locatees into the lumbering economy.

Beginning in about 1861, Ottawa Valley and small area lumber companies began buying the timber that the settlers were clearing from their free grant land. The majority of such transactions, a total of one hundred between 1861 and 1871, occurred in Galway, where white and red pine were cut and sold to a variety of buyers as complete pieces or as sawlogs. The Gilmour Company was active here, purchasing 885 white pine during the first five years of the decade.[37]

Local sawyers were also active. John Hunter's sawmill at Kinmount on the Galway–Somerville border was powered by the Burnt River's current. It was a cornerstone of the area's early settlement, supplying sawn lumber for home and road construction. Hunter bought timber from surrounding lots in Galway, Somerville, and Lutterworth. In 1862 he took 253 white pine from two of his own lots in Somerville. Between April and July 1862 he also purchased timber from Lutterworth's original free grant locates: Thomas White (151 white pine), Samuel Pierson (104), James McGuire (60), Thomas Grogan (60), and James Pocock (98). Unfortunately the source documents do not indicate the value of the timber or the rate at which it was purchased for any of these transactions.[38]

Boyd also used this system. For example, in 1862 he purchased over 900 sawlogs from one Lutterworth man. Boyd's business savvy was evident in other dealings as well when, for example, he integrated smaller operators into his company. Two individuals who had actively pursued locally cut timber, James Devlin and R. C. Smith, combined their companies with Boyd's in 1864 and 1871 respectively. By combining with Smith, Boyd pushed further north into Stanhope, buying 2,673 sawlogs at $.15 per piece during the summer of 1871. Such moves enabled Boyd to blaze toward the Burnt River and Gull River valleys.[39] As early as the winter of 1860-61, nearly 3,000 white pine were felled in Somerville alone.[40]

Boyd's decision to embrace the Bobcaygeon project paid off. The road extended into what is today Haliburton County, opening better winter access to lands previously shrouded by pine forests.

Settlers also drew a seasonal income from road work. As with Boyd, the free grant locatees' fortunes were linked to the road's success, and they had a vested interest in extending it as far north as possible. In January 1863 the road's northern agent reported that sixty-four locatees had extended settlement seventy miles north of Bobcaygeon to meet the Muskoka River. Some of them requested lots at the northernmost boundary so that in the future they might supply produce for Georgian Bay and Parry Sound markets to the west.[41] Farmers must have realized that with a distance of up to 100 miles, they were not likely to access the southern markets in Peterborough and Lindsay.

As settlement pushed north, locatees found a means to accelerate the road's construction and draw an income. Beginning in the autumn of 1866, many heads of the original families or their sons or nephews turned to road labour. The majority of men (at least fourteen) came from the most distant Township of Galway, although in 1866 – 67 all eight townships were represented. In total 142 men and boys engaged in road work.[42]

A sampling of those from each township who worked on the project in 1866-67 reveals a cross-section of economic situations. Farmers of small- and medium-sized lands, livestock raisers, and woodsmen entered into road work. The operation took place between May and December of 1866 and between July and October of 1867, drawing men during both the planting and harvesting seasons in both years. Manual labourers earned $1.00 per day; teamsters using their own draft animals received $1.50 daily.[43]

Robert Purdy, a native of Ireland and an original free grant locatee in Galway Township, emphasized the more extreme situation among the labourers. By 1862 he had chopped thirty-eight acres, yet had not planted a single crop. The family was evidently surviving on Purdy's work as a chopper and exchanging cash for food. The Purdys' farming fortunes had improved by 1864, but it is likely that even at age sixty-four Purdy continued his Canadian frontier experience as a manual labourer. He earned $100 on seven projects between May 1866 and October 1867.[44]

John Coulter was an atypical participant. The twenty-four-year-old Irish native arrived in Galway Township in 1857 accompanied by three family members. Remarkably, he had chopped over 684 acres by 1862 while maintaining a sufficient crop of wheat, oats, hay, potatoes, and turnips. During the autumn of 1866 he used his oxen and his own labour to earn $11.00 from road work. Coulter never again went back to the project; he had either found the experience monetarily unattractive, or perhaps he felt that his effort was better expended on his own property.[45]

The case of Lutterworth's Thomas Milburn also illustrates a popular practice along the Bobcaygeon Road. It had been hoped that once free grant locatees began breaking the land they would write their friends and relatives in the British Isles and encourage them to emigrate. Milburn arrived around 1862 and according to agent Hughes' report, resided with another person on lot 31. Within the next year they were joined by eight others, including Leonard and Robert Milburn, who were presumably brothers or otherwise related.[46] All three men engaged in road work, and during the 1870s and 1880s Thomas and Robert cut timber for the Boyd Company. Along with other area farmers, Robert sold hay to the company for the horses and oxen.[47]

Others supplied the project with food and draft animals. Minden's William Jervis, for instance, in part provided for his family of six during the 1866 – 67 season by hiring out his oxen. Although it is likely that some farmers sold surplus produce to the construction camps, Thomas Flemming's is the only recorded instance. Until late 1866 he was a labourer, but in December he sold $19.20 worth of potatoes and hay to the road labour camp at Trading Lake.[48]

There were two principal reasons for road labour's attractiveness to settlers: its proximity to the homestead and hard cash to supplement farmers' incomes in an area of limited commercial potential and outlets. In this respect, the relationship between settlement and road construction differed from other North American wooded frontiers, as very few locatees were drawn to lumbering. In the case of Mossom Boyd's operations, for example, settlers' names or those of relatives were conspicuously absent from pay lists. Most often Boyd hired experienced French-Canadian lumberjacks who had honed their skills in Ottawa Valley forests.

Local Knowledge: Road Construction and the Folly of Settlement

As settlement progressed along the Bobcaygeon and other roads, the lumbermen's encouragement of settlement became qualified. Specifically, lumbermen began to regret the encroachments of those whom they considered to be "pretending" settlers: those who deliberately chose marginal agricultural lands that also contained valuable pine forest.

In the rush to open the roads and locate people, the Crown Lands Department had neglected several important considerations. Essentially, the department had never specified at which point settlers could begin felling pine: could it be done during the initial year of occupation, or should it instead occur after five years of actual residence, improvement, and the granting of the title? Nor had the department rationalized its policies of timber licensing and free granting. Thus, it was possible for a free grant locatee to be assigned agriculturally unproductive land containing valuable pine stands.

By surveying and throwing open for settlement the unproductive lots that contained pine, contended the lumbermen, the province was encouraging the destruction of the forest. The free grant system held much promise, especially as it affected the lumbermen's operations. The pretending settler, however, sought and moved onto a marginal farming lot. He then harvested and sold the pine to the highest bidder and later moved off the land. Thus it was possible for the lumbermen to hold legal claim to the timber but never see a return on investment. In addition, the culled landscape and its littered forest floor invited fire, which portended disaster for both lumbermen and settlers.

Crown Lands timber agent James H. Burke of Bytown outlined the problem before a parliamentary committee in the early 1850s, before the roads were actually begun. While Burke saw merit in the free granting of what were considered wastelands, deemed unproductive from a logging standpoint, he viewed with trepidation the encouragement of settlement onto pine tracts that were traditionally the lumberman's domain. "The present system of [freely granting wastelands] has the tendency to conserve the pine timber," he explained, "as it spread[s] a local market for the produce of the backwoods settler over the longest space of time, without which the settlement of several hundred square miles of [the province's] best territory can scarcely be made."[49]

Burke's words are testament to the idea that Victorian-era agents could redesign the landscape to accommodate the prevailing economy. "We have an immense fertile territory stretching westward from Bytown to Lake Huron, and north-westward from Nipissing to Lake of the Woods," he asserted, "but our territory is a wilderness. In the center of the country named lies [*sic*] the timber fields of the Ottawa, at present yielding their first crop, which goes to build up the cities of the east and west. Nature has so arranged it, that this pine-producing territory does not possess a fertile soil.... This pine territory has its allotted end, and will subserve; perchance beneath those far-stretching forests repose rich mines of metal to tempt man's arm to delve the earth when the dark green canopy, which shuts out sunlight, has disappeared."[50] Burke continued, "[S]urrounding this pine territory and contiguous to the great lumber fields, is the large area to which we have alluded, possessing a fertile soil and timbered with hardwood. This timber has not the commercial value, and its destruction is not a national loss. This land is destined to sustain a large body of agriculturalists in close proximity to the great timber making centres."[51] As if to complement the forest activity, would-be farmers could cultivate the land and strategically allow for the raising of "grain, fodder and provisions, consumed in timber making, from eighty to ninety miles nearer the ground of consumption than we now do. While the lumber trade flourishes in pristine vigor population should be introduced."[52]

However, Burke did not want his plan to be misconstrued, for his was a plan that designated proper activities on suitable lands. As he noted, "let us not be understood to encourage the wanton, foolish and insane policy of the Crown Lands Department in surveying a township where nothing but pine and rock exist, or where to get a thousand acres of habitable land, settlers may be thrown in to spread fire and havoc through the pine forests; we go for keeping a fair line of separation between the lumbering and agricultural regions, as nature has laid it down."[53]

Along with the promise of land and the potential for commercial farming, the lumbering economy had helped to integrate settlement into the once exclusive domain of the lumberers. Seemingly the timber agents and lumbermen wanted the best of both worlds: winter access roads, nearby farmers and sawyers, and a reserve army of labour to accommodate the camps. If Burke's comments can be taken as representative, however, field agents (and in no small way the lumbermen) sought clearly defined settlement boundaries. They welcomed *settlers*, but they reservedly embraced *settlement*, especially if it was encouraged amidst the pine tracts.

By 1863 the conflict between lumbermen and pretending settlers had increased. Commencing in April 1863, a parliamentary committee heard testimony from lumbermen and others regarding the colonization road settlements. The lumberers' grievance was with pretending settlers and the Crown Land policies that encouraged them. They did not object to the local exchange conducted with actual settlers (as discussed above).

Alexander Dennistoun, lumber manufacturer at Fenelon Falls recounted his frustration with this policy. He related how in 1861 he purchased a license to cut timber on the unsold lots in Bexley Township. He understood that these lands would not be put on the market until he had had ample time to take the timber. Much to his surprise, when he attended another land auction in Lindsay three months later, the lots covered by his license were being advertised for sale by the Crown Lands agent at the remarkably low price of four shillings per acre. Dennistoun suffered a heavy loss; with settlement commencing he was able to take only about one hundred pieces of timber.[54]

Another lumberer, Ottawa's Allan Gilmour, went further in deriding the department's settlement approach. He questioned the very wisdom of the colonization roads, citing personal observation and geological data to underscore claims about the poor soils and frequent rock encountered by would-be farmers. Moreover, he was suspicious of the surveyors' reports that stated the opposite view. He even suggested that such surveys were conducted so as to keep the surveyors permanently employed in the field.[55]

Peterborough farmer Robert Strickland was familiar with the problem as well. He believed that the lumberer and the settler benefited one another, but only so long as the pine lasted. Once the timber was removed, the land was left in a rough state, and before the settler could put in a crop, Canadian thistle and other weeds took over the space where the pine had once grown. Should the settler then choose to burn the area and kill the weeds, he ran the risk of destroying other species that might otherwise be used for firewood or for home building. Strickland also attempted to explain, though not defend, the motivations of the so-called pretending settler. He explained that it was not always this character's desire to simply move onto a lot, clear the timber within one season, sell the timber to the highest bidder, and then move on. Some had a more long-range goal. A settler might, he explained, take possession of the lot "to prevent the lumber merchant from slashing over his bush, and by preserving this timber gives employment during the winter months for his team and himself, by conveying square timber or sawlogs to market, which enables him to get lumber to make such buildings as he may require on his farm."[56]

Strickland proposed that the best way to end the feud between the two groups was to plan the township more logically. Once assessed, the lands lacking agricultural potential should be reserved exclusively for the lumberers; the settlers could then take the remainder. "If this system were adopted," he argued, "the lumberer would save the ground-rent he now pays for lakes [and other portions upon which no pine grows] ... [and in so doing] the settler [would] have full control over such lands that were selected for him. You would find by this system that a much better class of settlers would go back to the new townships. There are very few good substantial men that will settle on land unless they can call it strictly their own."[57]

Perhaps Ezra Stephens spoke with the most authoritative voice. He summarized the matrix in which the settlement scheme occurred. Having been a Northumberland farmer for thirty-eight years, Stephens contemplated a move north along both the Hastings and the Addington roads. Upon hearing the much vaunted claims of the agents regarding superior soils, he followed his acquaintances north to investigate. During an 1861 visit along the Hastings Road, "the object of my visit," he recounted, "was not realized, for I found many of the settlers dissatisfied and discontented, the land poor, and the greatest part of it entirely unfit for settlement."[58]

The following year Stephens traversed the Addington Road and made enquiries of the settlers. He discovered that many emigrants from the old country, not being "competent judges" of the Canadian wild lands, had been deceived by the glowing reports of government surveyors and agents that had circulated in Europe. Desperate for the free grants of land, many had arrived during winter when snow blanketed their holding; come spring, when their hardwood trees were cleared off and the land burned over, those hardwood ridges were "then seen to be beds of rock and stone."[59] Those who had not already abandoned their lands, added Stephens, would do so at the first opportunity. Stephens also found that some had purposely "taken up grants for pine timber, knowing at the same time, that the land was unfit for settlement or agricultural purposes," and thus confirming the worst fears of critics of the colonization scheme.[60]

More damning than Stephens' description of life on the colonization roads was his contention that the Crown Lands agents' reports of crop yields were embellished. An agent visited the individual farms annually to ascertain the value of the wheat crop, even before the crop had been thrashed. Settlers, charged Stephens, were led to believe that if they gave a good account of their property, they would soon obtain their deeds. Thus the exaggerated, Eden-like accounts of boundless agricultural potential only became compounded.[61]

Environmentally, Stephens anticipated by fifty years the conclusions of a Commission of Conservation survey. The road openings and free granting system, he believed, had "the tendency to destroy the forests in more ways than one, as many settlers take up lands thickly wooded for the sake of the pine timber."[62] Often in attempting to burn off the cleared land, settlers allowed the fire to get beyond their control until it ran over large tracts of land that were unfit for settlement but that contained valuable and as yet uncut pine timber, thus destroying the stands.[63]

Although these comments refer to roads other than the Bobcaygeon, similar cases were unfolding there as well. During the 1860s several Minden residents, unable to maintain a farm or livelihood, requested alternative land in the newly opened townships of the province's southwest.[64]

For social, economic, and environmental reasons the Crown Land policies were a failure. Notwithstanding this testimony, however, the local press supported the colonization project. "We have just learned, with surprise and regret," expressed the *Peterborough Examiner* in March 1863, "that the Government are [*sic*] about stopping the opening up of the lands in the rear of this and the neighboring County.... [With Ottawa lumberman Allan Gilmour urging an end to colonization, it will] induce the Government to discontinue its [road building] operations up here, where so many settlers are constantly going in, and where there is so much good arable land.... Mr. Gilmour and the whole tribe of lumbermen are extremely anxious to keep as much land as possible in its primitive state, or rather they do not wish to see settlement increase."[65] Perhaps the local press viewed with suspicion the lumbermen's motivation in attempting to close the public out of the Crown Lands, or perhaps it did not fully accept that the project was failing on a number of counts; in any case, the local press only saw merit in the continuation of the plan.

By 1870 Boyd privately shared the other lumbermen's opinion and wished to keep settlement from interfering with his operations. So as to further detach his operations from the settlers, he instructed his Burnt River foreman to clear land and plant potatoes, turnips, and oats, which could be used as horse and oxen fodder. In doing so Boyd avoided purchasing such crops from area farmers.[66]

The settlement that arose along the Bobcaygeon road can best be characterized as incremental, while the relationship between settlers and the government's road was symbiotic. Locatees turned to road labour for cash supplements to their annual income; casual labour enabled them to remain in this low-yield farming area. The province relied upon local labour to earnestly push the road north.

Was the Bobcaygeon colonization road a success? From a social standpoint the project failed to attract a permanent agricultural community along the Shield. A later Commission of Conservation survey emphasized that population in the townships discussed had declined between 1901 and 1911. In 1901, 5,793 persons resided in the original townships, and by 1911 only 4,852 remained: a difference of 941 people.[67]

Ostensibly these townships were intended as farming communities. Although some pioneer families were probably self-sufficient to a degree, this level was reached only after much hard work. Contemporaries remarked that the land along the Bobcaygeon Road passed uneven and rocky areas of limestone and granite.[68]

Another commentator was more colourful in describing his first impressions of the area. George Thompson recalled his stagecoach trip along the road, probably in the early 1870s. "The Bobcaygeon road appeared to me to be one long drawn out tavern, for nearly every other house along it sold whiskey or beer, and without a license at that."[69] It was also regarded as a rough road and on that trip Thompson believed that his "toe nails would be shaken off."[70] Nearly all the settlers along the road were old soldiers, pensioners to whom the government had given grants of land for past good conduct and service. This fact disturbed Thompson and dissuaded him from "ever wanting to be a British soldier, for I thought if that was the way they rewarded those who had merited reward for good conduct I wondered what the fate of those could be who had bad conduct served up against them."[71]

The potential for extensive agriculture was obviously doubtful given the soil's thin and rocky composition. Thompson worked in the area's lumber camps for several years and remembered local farmers' comments. "Old pensioners say that they wished they had brought some of the old cannon captured in the Crimea war with them, so they might shoot the seed into the ground, for they said that was the only way they knew the seed would be successfully planted in that kind of soil."[72] Once duped by the government agents into settling on the Shield, farmers tried the same trick on unsuspecting newcomers. Thompson recalled that along the

Bobcaygeon Road a farmer, "when pointing out the good features of his farm probably to some stranger with a view of selling it, would always claim that the back fifty was splendid farming land; of course the settler could not help but admit that the front fifty was a little rough and rocky, for the stranger could usually see that for himself. This 'back' fifty racket got to be a well known remark, and it has often provoked a smile from parties who were not so green as they looked."[73]

J. I. Little's study of Québec's frontier settlement presents the best study with which to compare the Bobcaygeon Road project.[74] In the Upper St. Francis District during the nineteenth century, the state, church, and capital worked at various levels and under various guises to stem the tide of French-Canadian out-migration to New England mill towns. In time, and with mixed results, these vested interests achieved their goals. Many of those who ventured out along the colonization roads, however, found themselves on the fringes of poverty.

In English-speaking Canada West, on the other hand, administrators hoped to attract new immigrants away from Canada East and thus gain political clout through an expanded population. Once the mostly Anglo-Celtic emigrants had arrived, the state hoped to prevent them and native-born Canadians from moving to the United States. Colonization roads cut through Crown Lands appeared to be a good strategy with which to achieve these goals. Moreover, the project would satisfy the needs of lumbermen, who benefited from improved access, a nearby supply of farm produce, and a seasonal labour force. A permanent farming community was not possible along the Bobcaygeon Road, due mostly to the realities of the physical landscape.

Environmentally the project failed miserably. The road was intended to inspire an agricultural settlement, but it soon became evident that Crown Lands agents and some surveyors had painted an overly optimistic picture of the area's potential. Left with no other means of survival, settlers turned to the forest for income, thus upsetting the theorized balance between pioneer and lumberman. According to the Crown Lands administrators, settlers would aid the lumbermen, not compete with them. What ensued, however, was a race to clear the pine forest and reap a profit.

John Langton, a long-time Trent Valley resident and farmer, a one-time lumberman, and eventually Canada's public auditor, critically reviewed the process of forest frontier settlement at the time of the colonization project. In 1862 he cautioned that Canada's forest resources were finite and should be seen as such. For years he had witnessed the loss of forest cover due to settlement, fires, and wasteful cutting techniques. His first and most important suggestion for remedying these problems was to discourage indiscriminate settlement in the timber districts. "When the land is of such a quality as to support an agricultural population," asserted Langton, "by all means let the settlement proceed as rapidly as possible, for with all my respect for timber, I value a man more than a great many trees. But, a very large portion of our back country never can maintain a healthy settlement."[75] He was "convinced that a large portion of the new settlements in the back country, lying north of Lake Ontario, between the Ottawa and Lake Huron, will prove utterly worthless as the site of an agricultural population, but the attempt to form one will in the meantime have destroyed a mine of future wealth for Canada. The new lands ought to be classified, and only such as are really good should be open for indiscriminate settlement, and the rest ought to be reserved for lumbering purposes."[76] Langton's concern speaks to a larger North American idea embodied in the conservation movement: in fact his argument anticipated such debate by more than a generation.

Postscript:
Toward an Era of Conservation

IN 1913 the doyen of Canada's forestry educators, Bernard Fernow, reflected harshly on the condition of the upper Trent environment.[77] He contended that through fifty years of continuous culling the forest had acquired an unfortunate human imprint. The pine timber was now practically gone, and the species that did remain were of considerably diminished value. Wasteful selection of white and red pine and a disregard for other species had left a forest floor littered with felled trunks and decaying underbrush. Such a situation gave rise to repeated sweeping fires, resulting in a desert-like landscape of barren rock. The already thin soil was left with little defence against heavy rains, leaving topsoil vulnerable to being carried away by melting snow and spring rains.

In locally situating the problem, Fernow deduced that the Trent Canal, begun in the 1830s but as yet unfinished, would eventually suffer due to diminished water supply from the upper lakes. The human costs were high as well. Descendants of the original settlers were now few in number, their ranks thinned because of the marginal agricultural potential of the land and unrealized commercial connections to the west. Immediately north of Peterborough lay Methuen, Anstruther, and

Figure 9: Commission of Conservation, *Typical upland farm and farm buildings in Canada*, 1913; *Trent Watershed Survey: A Reconnaissance*, C.D. Howe and J.H. White, comp., with an introduction by B.E. Fernow (Toronto: Bryant Press), 96 – 97.

Burleigh Townships, whose forests had teemed with activity twenty-five to fifty years earlier. Here alone nearly 150,000 acres of desert-like conditions prevailed by the early twentieth century.

Fernow lay the blame for this environmental catastrophe squarely at the feet of the lumbermen and the provincial timber policy that had inspired their plundering of the pine forests. In the pre-Confederation period, provincial administrators had taken the concerns and demands of lumbermen seriously. Through their payment of timber licenses, valuable revenue was amassed and then channelled into other economic ventures, such as canal and road construction. Lumbermen enhanced the national economy by creating countless jobs at all levels of the lumbering process. Unfortunately, this axiom did not ring true along the colonization roads.

The area above Lakefield and Bobcaygeon was the worst example of the lumberers' excess. The descendants of the first back township settlers were now in dire straits. The land had been made unproductive and uninhabitable, claimed Fernow. "Not only have many farms been abandoned by the removal of their occupants to more hopeful conditions," he lamented, "but a considerable number that ought to be abandoned remain occupied by those who lack the means and energy to move, thus forming a poverty-stricken community."[78] He called for an enlightened stewardship policy that would address the natural and human condition. Given their reliance upon the forest, it is not surprising that the first Canadian proponents for forest management were lumbermen. A later generation would identify with Fernow's ideas and would urge the establishment of federally monitored reserves, wherein human settlement was strictly regulated. Concern over natural resource depletion was continent-wide and reached its zenith between approximately 1890 and 1930.[79]

Conclusion

THE example of the Bobcaygeon colonization road, like the project intended for Ops Township, illustrates the cognitive gap between first nature and second nature. Local knowledge and experience informed the rising opinion that colonization on the lumberman's frontier was a monumental task, yet only after more than a decade of abuse did this become obvious.

In part, the taking of the Trent Valley's forest can be understood as part of W. L. Morton's pointed assertion about the Victorian Era in Canada. "Here was nothing, or little, that was subtle or refined," he writes, "[b]ut here was the conscious, deft and masterful creation of human environment, the transforming of a wilderness, often harsh, often sparse, into a cultivated landscape, a human abode.... [T]o create a human habitat, to make the land habitable, and responsive to human needs, that might be articulate and sensitive, was to work a gigantic masterpiece, to draft a very shield of Achilles."[80]

The chimera that the Shield could be conquered scarred the upper Trent Valley and shattered the Victorian-era belief that habitation could be extended north. The state's involvement in the building of such transportation systems is by now a familiar theme. Road building is most closely associated with the other large infrastructure project, the Trent Canal, yet it also has meaning when considered as part of the development of the valley.

In the Trent Valley the achievement of these transportation objectives meant the reshaping of the landscape to facilitate such manifold human activities as agriculture and lumbering. Throughout the pre-Confederation period, the valley bioregion became a home to newcomers and a place that proved adaptable at times (but not always). This antiseptic summation is not to suggest that a more cerebral and genteel chronicle of the Trent Valley cannot be found. Indeed, it exists in the rich interpretive data left to us by the Trent Valley's foremost observer of environmental change, Catharine Parr Traill.

6

The Trent Valley Oracle: Catharine Parr Traill

Thus far, this has been a narrative of change and adaptation of various peoples—Aboriginal, European, Canadian—to the Trent Valley environment. Most often these actors sought home places and a certain reciprocity with the natural environment, as was the case with the Mississauga and the Anglo-Celtic settlers. At other junctures, such as in Ops Township or along the Bobcaygeon Road, we have encountered a vast cleavage between actual capabilities of the land (first nature) and the template of production that newcomers sought to impose upon it (second nature).

It is now useful to reflect upon another contemporary voice, that of Catharine Parr Traill, and her quest for a "place-based" narrative inspired by the very changes that we have so far considered. Indeed, Traill's residence in the Trent Valley during the nineteenth century coincides with the interactions that humans visited upon this bioregion. Moreover, Traill brings a unique female perspective into what has so far been an overwhelmingly male story. Her own understanding of natural history came from reading Gilbert White, the English cleric who defined the natural history essay for generations of English readers. Unlike White, however, Traill sought to introduce a historical dialectic and human agency into these observations in ways similar to the American commentator George Perkins Marsh.

The fictional and nonfictional works of Catharine Parr Traill are thus remarkable in that they underscore Old World ideas about nature and map their diffusion and transformation in the New. She fancied herself an amateur scientist, naturalist, and writer who hoped that Canadians would regard her work alongside that of White. In fact, she hoped to be remembered as the "Canadian Gilbert White."[1]

Above all, however, what emanates from her writing is a desire for what Peter Berg and Raymond Dasman refer to as a "reinhabitation" of the land. By this they mean "learning to live-in-place in an area that has been disrupted and injured through past exploitation [which] involves becoming native to a place through becoming aware of the particular ecological relationships that operate within and around it. It means understanding activities and evolving social behaviour that will enrich the life of that place, restore its life-supporting systems, and establish an ecologically and socially sustainable pattern within it."[2] Traill was indeed the Trent Valley oracle, providing a female reflection upon place and what it meant to arrive, settle, and mature in this ecological locale. She hoped to educate her contemporaries about this evolving home place, inducing them to see it in more holistic terms.

In addition to the frame of reference offered by Berg and Dasman, we might also understand Traill within the purview of British and European women who possessed an enthusiasm for nature study and who emigrated to North America during the nineteenth century. As Vera Norwood suggests, these women brought with them as cultural baggage a scientific knowledge, a romantic aesthetic, and gender-based behaviour codes, which then informed North American women as they developed their own voices for describing, valuing, and protecting the plants and animals they discovered.[3]

Catharine Parr Strickland was born in London in 1802, married a British officer, Thomas Traill, in 1832, and in that same year emigrated to Upper Canada. During her nearly seventy years in Canada she wrote over twenty-five books and articles, and at her death at age ninety-seven in 1899 she was celebrated as one of Canada's foremost nature writers.

However, Traill had more to relate to her audience than mere accounts of pleasant walks through the forest and the joys of gardening. Her writing presents a chronicle of the changes in the physical and cultural landscape in the Trent Valley that not only conveys a sense of her contemporaries' place in history but also parallels the transition of the Trent Valley from a wilderness to a frontier community during the nineteenth century. Whether visiting nearby Mississauga reserves, describing the state of newly arrived Anglo-Celtic immigrants, commenting on the construction of the Trent Canal, or recalling with ambivalence the march of civilization, Traill's writing is investigative, sometimes anecdotal, and always telling.

Scientific thought in nineteenth-century Canada has been explored by a small number of historians. Their studies have emphasized the colonial scientific heritage, the growth of institutions, and the role of prominent individuals in disseminating popular theories. Most often the practitioners examined were professional men, who usually engaged in rigorous debates in university lecture halls or conducted laboratory and field research. Carl Berger has outlined the study of natural history in Victorian Canada as both a professional and an amateur endeavour. William Logan's 1842 Geological Survey of Canada may be seen as one of the more important examples of the former; instances of the amateur form are sometimes difficult to characterize. To the middle-class person of that era, science was not an obscure diversion. Social gatherings were arranged around excursions into the wilderness and were recognized as a principled pursuit of consequence on the part of an educated, if not professional, scientific public. Berger acknowledges the integral contributions that nonprofessional naturalists such as Traill made to early Canadian science through the collecting and cataloguing of new types of flora and fauna. Suzanne Zeller notes that Traill's work was so highly regarded that the University of Edinburgh requested native botanical specimens from her; indeed, the 1862 London exhibition showcased her collection.[4]

Natural History: Catharine's Early Years

THE intellectual climate in which natural history study was conducted during the late eighteenth and nineteenth centuries was laden with controversy. Donald Worster has outlined the prevailing debates of this period and identifies two contradictory perspectives of the human–nature relationship: the imperial and the Arcadian. Both approaches subordinate nature to human use, but there are stark differences between them. The former stemmed from the Baconian desire to recover for humanity an esteemed position that was lost after the expulsion from the Garden of Eden. The Arcadian perspective, popularized by Gilbert White, advocated a stewardship of nature and studied the interrelatedness of species in what may be called an early ecological fashion. The commonality of these two visions was a firm Judeo-Christian belief that the earth was created by an omnipotent God.[5]

Traill turned to White as a font of knowledge. An Oxford-educated curate, White espoused an Arcadian worldview in his English country village. In *The Natural History of Selborne* (1789), the reader is taken through the countryside examining the local flora and fauna and its interaction with various other species. White marvelled, for example, that the waste of one species supplied food for others in perpetuity, as in the case of cattle, whose manure gave sustenance to insects, which were then consumed by fish, and so on. To White, nature was a great economist, efficient in every divinely created way; each element of the Selborne microcosm contributed in a meaningful way to the overall health of the larger living community. Though ecological in his approach, White was also utilitarian: he believed that through careful observation nature could be managed to better support humanity.[6]

White and the Arcadian tradition were revived and reintroduced into contemporary discourse during the Romantic era of the late eighteenth and early nineteenth centuries. William Wordsworth and others celebrated the English landscape and regarded nature highly in a way that paid homage to the Selborne curate. English Romanticism informed Anglo travellers and settlers alike as they encountered the Canadian wilderness. They described what they saw as "sublime" (the awe-inspiring Niagara Falls) or as "picturesque" (the tranquillity of the Thousand Islands along the St. Lawrence River). Such connotations provided them with emotional and cultural definitions as they came to grips with a very different landscape than that of the British Isles. Thus, in addition to an Arcadian view of nature, Traill brought with her a Romantic vocabulary with which to narrate what she found in her adopted land.[7]

Residing at an Elizabethan mansion in Suffolk, the Strickland girls enjoyed a progressive education in literature, history, the classics, arithmetic, and especially science. This instruction was bolstered by their father, who took Catharine on fishing excursions, taught her the names of English wildflowers, and introduced her to Izaak Walton's *The Compleat Angler* (1653) and White's *The Natural History of Selborne*. Given this context it is not surprising that young Catharine was enamoured with nature: she later praised the purposefulness of botanical study by stating that "merely to load our memories with the learned names of trees, and plants, and flowers, is after all but a barren and unsatisfactory acquisition.... [W]hat can unfold to the female mind sources of purer, more intellectual, yet simple enjoyment than the floral world?"[8] This training would serve her well in the New World.

Figure 10: Unknown, *Road through a pine forest*, in Catharine Parr Traill, *The Backwoods of Canada* (London: Nattali and Bond, 1836), 93 – 95.

Second Nature:
Homing in the Backwoods

IN 1832 Catharine Parr Strickland married Thomas Traill, a half-pay officer in the British Army, and moved to the Newcastle District of Upper Canada. Already residing in the district were two of Catharine's siblings: a brother, Colonel Samuel Strickland, and a sister, Susanna (Strickland) Moodie, who would also become a prominent early Canadian writer.

Traill's first New World literary piece was *The Backwoods of Canada* (1836). Here she related her initial experiences in an empirical fashion, highlighting factual information drawn from personal impressions and experiences that revealed to prospective settlers the landscape and reality of Upper Canadian life. The book began as a collection of letters that Traill wrote to her mother in England, but as Marian Fowler notes, it became something of a true-to-life diary modelled on Daniel Defoe's fictional *Robinson Crusoe* (1719). Readers discovered the Canadian wilderness and the emerging communities; furthermore, it evinced the transition from Traill's pastoral English countryside to that of the rugged Canadian Shield.[9] On the bumpy wagon road from Cobourg north to Rice Lake, she described the land that "rises into bold sweeping hills and dales, and again descends to the valley.... Beyond this beautiful lake the land again rises in graceful sweeps northward."[10]

The forest understandably held Traill's attention throughout most of the book; settlers were surrounded by it and had to conquer it through chopping and burning in order to improve their lands. She was certainly cognizant of this fact, although she occasionally criticized the scale of forest clearance and lamented the loss of ornamental covering such as she had experienced in Britain. "On first coming to this country," she reflected in November 1833, "nothing surprised me more than the total absence of trees about the dwelling-houses and cleared land; the axe of the chopper relentlessly levels all before him. Man appears to contend with the trees of the forest as though they were his most obnoxious enemies; for he spares neither the young sapling in its greenness nor the ancient trunk in its lofty pride; he wages war against the forest with fire and steel."[11] She recognized the necessity of forest clearance, but her Arcadian and Romantic preconceptions prevented her from fully condoning the march of progress.

Her aesthetic sensibility was also offended by the lumberman's assault; following a burning, she recalled that she wished to preserve "a few pretty sapling beech-trees that pleased me.... I desired the choppers to spare them; but the only one that was saved from destruction in the chopping had to pass through a fiery ordeal, which quickly scorched and withered up its gay green leaves: it now stands a melancholy monument of the impossibility of preserving trees thus left.... The only thing to be done if you desire trees," she advised, "is to plant them while young in favourable situations, when they take deep root and spread forth branches the same as the trees in our [English] parks and hedgerows."[12] Thus, she attempted to impose an intellectual construct upon the Canadian landscape and strike a balance between aesthetic and functional woodlands reminiscent of the English countryside. In future, she hoped that on her family's homestead several forested acres could remain "in a convenient situation," with the old timber being chopped and drawn out for firewood and the younger growth left for adornment, so as to combine "the useful with the ornamental."[13]

The same urge to connect her new surroundings to familiar English ideas of landscape aesthetics and categorization is evidenced in Traill's attitude toward Upper Canadian flora and fauna. One of her first inclinations upon reaching the family lot was to assess the local flowers. "I have begun collecting," she noted, "and though

Figure 11: Unknown, *Newly cleared land*, in Catharine Parr Traill, *The Backwoods of Canada* (London: Nattali and Bond, 1836).

the season is far advanced, my *hortus siccus* boasts of several elegant specimens of ferns."[13] She also recorded the growth of yellow Canadian violets, Michaelmas daisies, shrubby asters, festoon pines, and other evergreens, and because"much of the botany of these unsettled portions of the country is unknown to the naturalist, and the plants are quite nameless, I take the liberty of bestowing names upon them according to inclination or fancy."[15] Traill rarely restrained this tendency toward botanical nomenclature. One year after settling, she supposed that"our scientific botanists in Britain would consider me very impertinent in bestowing names on the flowers and plants I meet with in these wild woods: I can only say, I am glad to discover the Canadian or even the Indian names if I can, and where they fail I consider myself free to become their floral godmother and give them names of my own choosing."[16]

Traill was consistently optimistic about life in Upper Canada, and the Trent Valley in particular. She expressed her support for the rumoured Trent Canal in the Peterborough area, which was remote and unconnected to much of the province due to poor roads. "This noble work," she wrote to a friend, "would prove of incalculable advantage, by opening a direct communication between Lake Huron and the inland townships at the back of [Lake] Ontario with the St Laurence [*sic*] [River]."[17] "Canada is a land of hope," She continued "[H]ere everything is new; everything is going forward; it is scarcely possible for arts, sciences, agriculture, manufactures, to retrograde; they must keep advancing; though in some situations the progress may seem slow, in others they are proportionately rapid."[18] She thus displayed a recognition of the necessity, indeed desirability, of material progress. Although Traill at times lamented the loss of forest, in her early years she looked wide-eyed upon the settlement trail being blazed through the province. This attitude would become tempered several years later, however, and largely rejected in her old age as she witnessed extreme changes in the Canadian landscape.

Traill's next major literary contribution was *Canadian Crusoes: A Tale of the Rice Lake Plains* (1852), in which she wove a tale of human adventure with natural history. The plot revolves around three teenagers lost in the forest, who, when forced to survive in the wild, learn to live off of the land as skilfully as natives. The story continues with encounters between the Mohawk and Chippewa, a commentary on their spiritual connection to nature and their hunting practices, and finally the return of the three young Canadians to their family.[19]

While crafting a popular novel accessible to a young audience, Traill also donned the role of historian. She regreted the human and material changes evident in the physical landscape, and at the novel's close the three protagonists reappear as adults, thus allowing Traill the narrator to reflect upon the altered environment. "What changes a few years make in places!" she remarked. "That spot over which the Indians roved, free of all control, is now a large and wide-spreading town [Peterborough]."[20] Spanning the Otonabee River is a great bridge, which the "townspeople and the country settlers view ... with pride and satisfaction, as a means of commerce and agricultural advantage"; now the "sound that falls upon the ear is not the rapids of the river, but the dash of mill wheels and mill dams, worked by the waters."[21] The lives of the local Mississauga had also changed; once they hunted and fished at such spots unhindered, but by the 1850s "the Indian only visits the town ... to receive his annual government presents ... [or to trade for goods which his] white brethren has [*sic*] made him consider necessary to his comforts, to supply wants which have now become indispensable."[22] "A new race is rising up," she lamented, "and the old hunter will soon become a being unknown in Canada."[23]

Traill astutely summarized the plight of Aboriginal peoples in Canada. In her view, they had become victims of history: once they were integral to the fur trade and served as allies of the French and British during the quest for imperial dominance over the northern half of North America. However, between the 1830s and 1850s they had become an impediment to progress, and colonial policy now addressed two prospects: the Aboriginals' removal from the path of Canadian settlement or their assimilation into Western "civilization." The irony was never lost on Traill, who saw this bleak state as a historical process that had resonance for both humans and the environment.[24]

The cultural transfer of Europe into North America signalled a theme in human history for Traill. Concern over the fate of Aboriginal peoples and a determination to integrate them into an account of local history are two of her literary trademarks. In the mid-nineteenth century they appeared to many observers as a people with an uncertain future, but for Traill they had a past that was relevant to the story of the Trent Valley; moreover, that past could be juxtaposed with European advancement and environmental change.

Other gestures by Traill also indicate respect for Aboriginal culture. At a time when middle-class settlers were attempting to furnish their homes in an English-Victorian style, the Traills' parlour floor was adorned by "a handsome Indian mat" that she had brought back from one of her many visits to her Mississauga neighbours.[25] In the aftermath of the 1885 Northwest Rebellion, she requested from her son William, a chief trader with the Hudson's Bay Company, any information about the Aboriginals of that area for use in a magazine article.[26] Much later, in 1897, at a meeting of the Peterborough Historical Society she presented a collection of artifacts given to her by her long-time "Hiawatha friends" from the nearby reserve of the same name; she hoped these artifacts would be displayed in the society's museum. Additionally, she took advantage of the ongoing construction of the Trent Canal to hunt for fossils on behalf of the museum.[27] Traill also championed the preservation of an island near her Lakefield home where a young Mississauga woman, Polly Cow, was purported to be buried. In the early 1890s Traill successfully petitioned the federal government to set the island apart from commercial or recreational use, and in recognition of her contributions to Canada she was awarded its ownership.[28] Finally, in the last year of her life, she took to closing a letter to her dear friend Sir Sanford Fleming with a phrase used by her Aboriginal women friends: "I bless You in my heart." For Traill, the past imparted its own wisdom, as well as clues that were worth recovering for posterity's sake.[29]

In *Canadian Crusoes* Traill moved toward a style that combined contemporary narrative with historical commentary, a literary approach that became more prominent in her later nature essays. She also broadly recounted environmental history by describing how the land and its original inhabitants had been changed, thus acknowledging that they too were actors in the drama of settlement. When she arrived in Upper Canada at age thirty, Traill regarded the progress toward permanent settlement as a positive endeavour in nearly every way. By 1852, at age fifty, this opinion had been tempered by the past twenty years in which she had witnessed in her own neighbourhood alone the marginalization of the Mississauga, the clearing and burning of large tracts of forest, and the construction of homes, roads, towns, and canals. Traill may have wondered if this marvellous progress came at a cost to the more natural features of the Trent Valley. As an amateur botanist and appreciator of nature, she began at mid-life to question the wisdom of an imperial outlook, which held the natural world and Aboriginal cultures as subaltern to colonizers.

The forest was also in peril. From her earliest days in her adopted country, Traill was astounded by the density of the forest; however, by the 1850s, such verdant scenes had been converted into agricultural communities, prompting her to intone:

> O wail for the forest, the proud stately forest
> No more its dark depths shall the hunter explore,
> For the bright golden main
> Shall wave free o'er the plain,
> O wail for the forest, its glories are o'er.[30]

Was Traill merely lamenting the loss of trees and their replacement with wheat plains in and of themselves, or was she also equating their passing with a sense of historical time? Clearly, it was the latter. In previous works she had acknowledged the dominance of Aboriginal peoples over the continent and the transition from those cultures' hegemony to that of the Europeans. Just as she saw Aboriginals and nature as actors in the process of changing landscapes in *Canadian Crusoes*, she recognized that Euro-American agency in the expansion of North American frontiers had consequences.

In "Ramblings By the River" (1853), Traill continued to suggest avenues for historical description. Reflecting on her first days in Canada more than thirty years before, Traill recalled how, without many books or other diversions, she became a "forest gleaner" out of necessity.[31] In Canada she was first struck by the colony's natural characteristics, and she could not resist comparing it to her native Britain. "The calm unruffled waters of England" were designed "as if by Nature to enrich and fertilize her soil.… [They] are unlike the wild streams of Canada."[32] On these untamed watercourses thousands of squared pine timber were annually driven to

Canadian ports and then sent to distant markets across the Atlantic. "Might not a history of no mean interest, be written of one of these massive timbers?"[33] Traill asked. The tale would begin with a pine cone's fall from atop the tree onto the ground, where it would then grow into a tall pine and then be chopped and shipped, ultimately reaching a British dock-yard. "Shall we believe," she pointedly asked, "that no providential care was extended over that seed which was in the course of time to undergo so many changes, and which might even be connected with the fate of hundreds of human beings?"[34] Again, Traill noted the forest's dominance, but more telling was her interest in nature as a force in human history.

Similar themes are embodied in her 1853 "A Walk to Railway Point." By the mid-nineteenth century, settlement, commercial agriculture, and lumbering necessitated the construction of roads and canals as Trent Valley inhabitants sought to become better integrated with Canadian cities. An ambitious scheme by railroad entrepreneurs to construct a bridge across Rice Lake and connect Cobourg to Peterborough was short-lived due to the structure's instability. Traill favoured such a plan, believing that it would greatly benefit the back townships and generally bring prosperity to the area. Yet with ever-recurring doubt, she also questioned both the intrusion of more machine technology into her surroundings and the apparent lack of empathy with which her Canadian contemporaries approached the ravage of such natural scenes. "As a lover of the picturesque," she wrote, "I must confess that I have a great dislike to railroads. I cannot help turning with regret from the bare idea of scenes of rich beauty being cut up and disfigured by these intersecting veins of wrought iron, spanning the beautifully old romantic hills and rivers of my native land; but here, in this new country, there is no such objection to be made, there are no feelings connected with early associations, to be rudely violated; no scenes that time has hallowed to be destroyed. Here, the railroads run through dense forests, where the footsteps of man have never been impressed."[35]

Thus, at mid-century Traill adopted a historical approach that sought to uncover the relationship(s) between humankind and nature. George Perkins Marsh would later employ a similar tact in *Man and Nature; Or, Physical Geography as Modified by Human Action* (1864), and is correctly recognized as the first environmental historian, but it is significant that another North American saw a kindred connection to the past. Traill's sense of history compares with that of Marsh, although the latter's writing is grounded in an American tradition of land use, and his contact with the Mediterranean region further broadened his international perspective, a quality that is absent in Traill's imperial-colonial Canadian approach.

Figure 12: *Portrait of Catharine Parr Traill.*
Archives of Ontario, s 2154.

Second Nature:
The "Canadian Gilbert White"

By her eighties, Traill's audience had become national and international. At this point she was writing in great detail about specific places within the Trent Valley. As she grew older, she appreciated the most minute qualities of nature as opposed to the sublime and picturesque panoramas that had first captured her imagination as a newcomer to Upper Canada.

Studies of Plant Life in Canada (1885) was Traill's first complete and systematic effort to botanize her neighbouring wild spaces. She regarded with importance each and every species, from weeds and mosses to shrubs and trees; with a keen eye for detail, she conducted a survey of the region's flora and commented upon habitat and medicinal and commercial uses. "As civilization extends through the Dominion," she noted, "and the cultivation of the tracts of the forest land and prairie, destroys the native trees and the plants that are sheltered by them, many of our beautiful wild flowers, shrubs and ferns will in the course of time, disappear from the face of the earth, and be forgotten, it seemed a pity that no record of their beauties and uses should not be preserved."[36] Providence, she asserted, cared for every species no matter how small. Even the carpetweeds outside her door had a place, and if God had "cared for the preservation of even the lowliest of the herbs and weeds[,] [w]ill He not also care for the creature made in His own image? Such are the teachings which Christ gave when on the earth. Such teachings are still taught by the flowers of the field."[37] Every plant, flower, and tree," Traill maintained, "has a simple history of its own ... [forming] a page in the great volume of Nature ... and without it there would be a blank—in nature there is no space left unoccupied."[38]

During Traill's sixty-seven years in Canada she observed many cultural and environmental changes, which were noted in her many publications. Her literary goals were to observe, catalogue, educate, instruct, and advise. As Marian Fowler notes, Traill was "the first Canadian to realize that naming the land is the first step towards taming it, and making it one's own."[39] Traill herself, in 1894, recalling the sight of her first sunset in Canada, remarked that "I wished to claim all the loveliness of Canada, the country of our adoption and henceforth our home."[40]

Traill's connection to the Trent Valley cannot be understated. During her residence there she witnessed firsthand the Anglo-Celtic encounter with nature.

For her there was meaning in the drama of adaptation and this emerged in her writing. Moreover, her corpus suggests reconstitutions of humankind's relationship to the Trent Valley environment, a reinhabitation that made nature relevant to her contemporaries. Her place-based expression provides a unique foil to other stories, such as the tragedy in Ops Township or the struggle of settlement along the Bobcaygeon Road, which were perforce inspired by Victorian ideas of progress. Ultimately, for Traill, the land was as much a factor, indeed an actor, in these events as the humans involved. Her life's quest was to celebrate this fact in the Trent Valley bioregion.

7

Conclusion

> The bee throws off her annual swarms.... It is possible that the young insects do not like to quit the hive or hollow the tree where they have been fed and nurtured. Yet parental oversight compels them to migrate to other trees and hives, not only for want of room, but because they must seek for honey in other fields. These unknown fields to the bee, with her confined vision, are the Canadas, Australias, and New Zealands of our world.
>
> Major Samuel Strickland,
> *Twenty-Seven Years in Canada West; Or, the Experience of an Early Settler* (1853)[1]

During the nineteenth century new frontiers were being explored throughout North America, while older ones, such as those of Aboriginal peoples, were becoming blurred or redefined. These new frontiers, including the Trent Valley bioregion, need to be seen as part of an international phenomenon involving the expansion of European peoples and ideas. Indeed, cultural and intellectual transport from Europe to North America is significant when considering Upper Canada's foundation. Like other regions of the continent, the province was not created in isolation.[2] It was instead the manifestation of European influence, which had reached the continent during the pre-industrial era. Indeed, such "settler societies" need to be analyzed in global terms, taking into account first nature and second nature contexts. This is precisely what I have striven for in relating the bioregional history of the Trent Valley.

In framing the wider portrait of global environmental change that the model of the Trent Valley suggests, we might usefully draw upon Thomas R. Dunlap's recent *Nature and the English Diaspora*. He persuasively argues that such "New" worlds as the Trent Valley were refashioned to resemble the familiar worlds of the British Isles.[3] Many physical challenges—dense forests, long winters, isolation—confronted the Old World travellers. By drawing upon their experience in the British Isles, they managed to overcome most of these impediments. In doing so, they replicated to some extent the Anglo-Celtic world that they had left: for example, Irish farming strategies, models of village life, the desire for a home place, and imperial and Arcadian worldviews were transmitted. The Trent Valley may have initially appeared as a New World, but the pioneers, like the mythical bees described by Major Strickland, quickly reinvented it as a familiar world in both a human and physical sense as they adapted to this bioregion.

Above all, the Trent Valley example offers a Canadian contribution to the study of bioregional history. According to Dan Flores, bioregionalism holds the promise of linking ecological locale and human cultures intimately within the same narrative.[4] Having done so in this study, we see that the task of adaptation to new lands was an ongoing process, and often one that yielded mixed results.

Studies such as this also offer a reflection on place. The search for and adaptation to a home place, whether by the Hopewell peoples or by Anglo-Celtic migrants, gives historical meaning to the Trent Valley's place within Canadian history. Though envisioned differently by each group, the common motivation was to make the land their home and to lay a foundation for future generations.

Seeing the interplay of humankind and the natural environment in this way serves to connect the Trent Valley to Ontario's past. Here comparisons to kindred studies by Thomas F. McIlwraith and J. David Wood are appropriate.[5] Each regards the Ontario landscape as a product of history, one that was shaped to fit particular ideas of what the land and resources meant to settlers and what opportunities they held for sustaining these early populations. Wood, in particular, highlights the fact that the first half of the nineteenth century was a period of tremendous ecological alteration, inspired in large part by the introduction of permanent agriculture. The case of the Trent Valley serves to validate his claim.

The upshot of such a bioregional analysis, ultimately, is that the land itself becomes a key actor in the narrative. It is an important ingredient in the histories that we weave about past places in Canada. Moreover, in better understanding the limitations of the earth, we might, as Berg and Dasman assert, reinhabit the earth and learn "to live-in-place in an area that has been disrupted and injured through past exploitation. It [is an exercise that] involves becoming aware of the particular ecological relationships that operate with and around it."[6] Viewing the natural world in such a way lends hope that as the new century dawns, we might begin to see ourselves and our relationships to the environment in healthier terms.

Notes

Chapter 1

1. Hugh MacLennan, *Seven Rivers of Canada* (Toronto: Macmillan of Canada, 1961), viii.
2. See, for example, the essays in Chad Gaffield and Pam Gaffield, eds., *Consuming Canada: Readings in Environmental History* (Toronto: Copp-Clark, 1995).
3. Edwin C. Guillet's coinage is not only appropriate, but practical. In his edited *The Valley of the Trent* (Toronto: Champlain Society, 1957), Guillet broadly defines the area that includes the Kawartha lakes and the Trent River Valley under this designation. I have taken the liberty of further compacting the label.
4. Thomas R. Dunlap, *Nature and the English Diaspora: Environment and History in the United States, Canada, Australia, and New Zealand* (Cambridge and New York: Cambridge University Press, 1999); Tom Griffith and Libby Robin, eds., *Ecology and Empire: Environmental History of Settler Societies* (Edinburgh: Keele University Press; Seattle: University of Washington Press, 1997). Of course, the foundational works on the topic of European expansion and its environmental change are Alfred W. Crosby, *The Columbian Exchange: Biological and Cultural Consequences of 1492* (Westport, CT: Greenwood Press, 1972) and *Ecological Imperialism: The Biological Expansion of Europe, 900 – 1900* (Cambridge and New York: Cambridge University Press, 1986); Carolyn Merchant, *Ecological Revolutions: Nature, Gender, and Science in New England* (Chapel Hill and London: University of North Carolina Press, 1989); Stephen J. Pyne, *Burning Bush: A Fire History of Australia* (New York: Henry Holt and Co., 1991); Richard H. Grove, *Green Imperialism: Colonial Expansion, Tropical Edens and the Origins of Environmentalism, 1600 – 1860* (Cambridge and New York: Cambridge University Press, 1995).
5. Carolyn Merchant, ed., *Major Problems in American Environmental History: Documents and Essays* (Lexington, MA: D.C. Heath and Co., 1993), Appendix, vii.
6. Dan Flores, "Place: An Argument for Bioregional History," *Environmental History Review* 18 (Winter 1994), 5.
7. William Cronon, *Nature's Metropolis: Chicago and the Great West* (New York: W. W. Norton, 1991), xix.
8. L. J. Chapman and D. F. Putnam, *The Physiography of Southern Ontario*, 2d ed. (Toronto: University of Toronto Press, for the Ontario Research Foundation, 1966), 13, 284.
9. Ibid., 22, 27.
10. Ibid., 99, 296 – 7.
11. Ibid., 313 – 6.
12. Ibid., 158; Guillet, ed., *The Valley of the Trent*, xxix.
13. Peter Adams, "The Climate of Peterborough and the Kawarthas:

Past and Present," in *Peterborough and the Kawarthas*, ed. Peter Adams and Colin Taylor (Peterborough, ON: Heritage Publications, 1985), 31; Fred Helleiner, "The Biogeography of the Peterborough Region," in *Peterborough and the Kawarthas*, 50; Jim Buttle, "The Kawartha Lakes," in *Peterborough and the Kawarthas*, 82, 92.

14. P. M. Catling, V. R. Catling, and S. M. McKay-Kuja, "The Extent, Floristic Composition and Maintenance of the Rice Lake Plains, Ontario, Based on Historical Records," *Canadian Field-Naturalist* 106 (January–March 1992), 83.
15. Guillet, ed., *The Valley of the Trent*, 5; James E. Anderson, *The Serpent Mounds Site Physical Anthropology*, Occasional Paper 11, Art and Archaeology, Royal Ontario Museum, University of Toronto (Toronto: University of Toronto Press, 1968); Catling and McKay-Kuja, "The Extent," 83.
16. Hermann Helmuth, *The Quakenbush Skeletons: Osteology and Culture*, Trent University Occasional Papers in Anthropology, no. 9 (Peterborough, ON: Trent University, 1993); Michael W. Spence and J. Russell Harper, *The Cameron's Point Site*, Occasional Paper 12, Art and Archaeology, Royal Ontario Museum, University of Toronto (Toronto: University of Toronto Press, 1968); Richard B. Johnston, *The Archaeology of the Serpent Mounds Site*, Occasional Paper 10, Art and Archaeology, Royal Ontario Museum, University of Toronto (Toronto: University of Toronto Press, 1968); Richard B. Johnston, *Archaeology of Rice Lake, Ontario*, Anthropology Papers, no. 19 (Ottawa: National Museum of Canada, 1968). See also William A. Ritchie, *An Archaeological Survey of the Trent Waterway in Ontario, Canada and its Significance for New York State Prehistory*, Research Records of the Rochester Museum of Arts and Sciences, no. 9 (Rochester: Rochester Museum of Arts and Sciences, 1949).
17. Joan M. Vastokas and Romas K. Vastokas, *Sacred Art of the Algonkians* (Peterborough, ON: Mansard Press, 1973), 27.
18. Catling and McKay-Kuja, "The Extent," 83.
19. Catharine Parr Traill, *The Backwoods of Canada: Being Letters from the Wife of an Emigrant Officer, Illustrative of the Domestic Economy of British America* (London: Charles Knight, 1836; reprinted, Toronto: McClelland and Stewart, 1989), 59. See also E. S. Dunlop, ed., *Our Forest Home: Being Extracts from the Correspondence of the Late Frances Stewart* (Toronto: Presbyterian Printing and Publishing Co., Ltd., 1889), 43 – 5.
20. Catling and McKay-Kuja, "The Extent," 82 – 3.
21. Isabel Skelton, *The Backwoodswoman: A Chroncile of Pioneer Home Life in Upper and Lower Canada* (Toronto: Ryerson Press, 1924), 188 – 91.
22. W. T. Macoun, "Why Our Field and Roadside Weeds are Introduced Species," *Ottawa Naturalist* 19, 6 (1905): 124 – 5; Charles MacNamara, "Champlain as a Naturalist," *Canadian Field-Naturalist* 40, 6 (1926): 125 – 33; and, R. M. Saunders, "The First Introduction of European Plants and Animals into Canada," *Canadian Historical Review* 16 (December 1935): 388 – 406.
23. See also Richard Stuart, "'History is concerned with more than ecology, but it is concerned with ecology too': Environmental History in Parks Canada," unpublished paper, Canadian Historical Association Meeting, Calgary 1994. Stuart and I have come to our conclusions separately from one another.
24. Carl Berger, *The Writing of Canadian History: Aspects of English-Canadian Historical Writing Since 1900*, 2d ed. (1976;

Toronto: University of Toronto Press, 1986), 117.

25. Harold A. Innis, *The Fur Trade in Canada: An Introduction to Canadian Economic History* (New Haven: Yale University Press, 1930); *The Cod Fisheries: The History of an International Economy* (New Haven: Yale University Press, 1940).
26. Arthur R. M. Lower, *The North American Assault on the Canadian Forest: A History of the Lumber Trade Between Canada and the United States* (Toronto: Ryerson Press, 1938); *Great Britain's Woodyard: British America and the Timber Trade, 1763 – 1867* (Montreal: McGill-Queen's University Press, 1973). In this connection see R. P. Gillis and Thomas Roach, *Lost Initiatives: Canada's Forest Industries, Forest Policy, and Forest Conservation* (New York: Greenwood Press, 1986).
27. David McNally, "Staple Theory as Commodity Fetishism: Marx, Innis, and Canadian Political Economy," *Studies in Political Economy* 6 (Autumn 1981): 35 – 63.
28. D. G. Creighton, *The Commercial Empire of the St. Lawrence, 1760–1850* (Toronto: Ryerson Press for the Carnegie Endowment for International Peace, 1937), 5.
29. Ibid., 2 – 3.
30. Ibid., 6 – 7.
31. W. L. Morton, "Clio in Canada: The Interpretation of Canadian History," *University of Toronto Quarterly* 15 (April 1946): 227 – 34.
32. W. L. Morton, *Manitoba: A History* (Toronto: University of Toronto Press, 1957).
33. W. L. Morton, "Victorian Canada," in *The Shield of Achilles: Aspects of Canada in the Victorian Age*, ed. W. L. Morton (Toronto: McClelland and Stewart, 1968), 311.
34. Ibid., 312.
35. Ibid., 330.
36. J.M.S. Careless, "Frontierism, Metropolitanism, and Canadian History," *Canadian Historical Review* 35 (March 1954): 1 – 21.
37. William Cronon, *Nature's Metropolis: Chicago and the Great West* (New York: W. W. Norton, 1991), 400 – in.
38. J.M.S. Careless, *Canada: A Story of Challenge*, rev. ed. (1953; Toronto: Macmillan of Canada, 1970), 3.
39. Andrew H. Clark, *The Invasion of New Zealand by People, Plants, and Animals: The South Island* (New Brunswick, NJ: Rutgers University Press, 1949).
40. Andrew H. Clark, *Three Centuries and the Island: A Historical Geography of Settlement and Agriculture in Prince Edward Island* (Toronto: University of Toronto Press, 1959).
41. Andrew H. Clark, *Acadia: The Geography of Nova Scotia to 1760* (Madison: University of Wisconsin Press, 1968).
42. Ibid., 392.
43. Ramsay Cook, "Canada: An Environment Without a History?" unpublished paper, "Themes and Issues in North American Environmental History," University of Toronto, 24 – 26 April 1998. See also his "Cabbages Not Kings: Towards an Ecological Interpretation of Early Canadian History," *Journal of Canadian Studies* 24 (Winter 1990 – 91): 5 – 16.
44. Ramsay Cook, "Canadian Centennial Celebrations," *International Journal* 22 (Autumn 1967): 659 – 63; and, J.M.S. Careless, "'Limited Identities' in Canada," *Canadian Historical Review* 50 (March 1969): 1 – 10.
45. Cole Harris, "Regionalism and the Canadian Archipelago," in *Heartland and Hinterland: A Regional Geography of Canada*, 3d. ed., ed. Larry McCann and Angus Gunn (Scarborough, ON: Prentice-Hall Canada Inc., 1998), 395.
46. Ibid., 402.

Chapter 2

1. Peter Russell, "Proclamation to Protect the Fishing Places and the Burying Grounds of the Mississagas [*sic*] (December 1797)," vol. 2, *The Correspondence of the Honourable Peter Russell, with Allied Documents Relating to His Administration of the Government of Upper Canada During the Official Term of Lieut.-Governor J. G. Simcoe While on Leave of Absence*, ed. E. A. Cruikshank and A. F. Hunter (Toronto: Ontario Historical Society, 1935), 41.
2. William Cronon, *Changes in the Land: Indians, Colonists, and the Ecology of New England* (New York: Hill and Wang, 1983).
3. A. F. Chamberlain, "Notes on the History, Customs and Beliefs of the Mississauga Indians," *Journal of American Folklore* 1 (1888), 155 – 6, quoted in Edward S. Rogers, "The Algonquian Farmers of Southern Ontario, 1830 – 1945," in *Aboriginal Ontario: Historical Perspectives on the First Nations*, ed. Edward S. Rogers and Donald B. Smith (Toronto: Dundurn Press, for the Ontario Historical Studies Series, 1994), 134.
4. Emma Jeffers Graham, "Three Years Among the Ojibways, 1857 – 1860," quoted in Edwin C. Guillet, ed., *The Valley of the Trent* (Toronto: Champlain Society, 1957), 19.
5. Ibid., 20.
6. Archives of Ontario (hereafter AO), Reginald Drayton Autobiography, MU949, series F, file 17, vol. 2, 1871 – 80 [typescript], 14.
7. Catharine Parr Traill, *The Backwoods of Canada: Being Letters from the Wife of an Emigrant Officer, Illustrative of the Domestic Economy of British America* (London: Charles Knight, 1836; reprinted, Toronto: McClelland and Stewart, 1989), 134 – 5.
8. Thomas Need, *Six Years in the Bush; Or, Extracts from the Journal of a Settler in Upper Canada* (London: Simpkin, Marshall and Co., 1838), 51, 68, 87 – 8.
9. Traill, *The Backwoods of Canada*, 132 – 4.
10. Robert J. Surtees, "Land Cessions, 1763 – 1830," in Rogers and Smith, eds., *Aboriginal Ontario*, 112.
11. Canada, vol. 1, *Indian Treaties and Surrenders. From 1680 to 1890* (Ottawa: Queen's Printer, 1891), no. 20, 48 – 9.
12. Cronon, *Changes in the Land*, 62 – 7.
13. National Archives of Canada (hereafter NA), Record Group 5, Series A1, Civil Secretary's Correspondence, Upper Canada Sundries, vol. 51, 25477 – 80, Charles Hayes to George Hillier, 6 March 1821 (reel C-4606). I thank Professor Bryan D. Palmer for this reference.
14. NA, RG5, A1, vol. 52, 26050 – 3, Hayes to Hillier, 23 May 1821 (C-4606).
15. *Journal of the House of Assembly of Upper Canada*, 1829 (hereafter *JHA*), 22, 23.
16. AO, RG1, series A-I-6, Crown Lands Department Correspondence, vol. 8, envelope 6, 7788 – 9, Peter Jones to Peter Robinson, 28 Sept. 1829 (MS 563/reel 8).
17. *JHA*, 1830, 21, 24, 25, 75.
18. *JHA*, 1831, 34, 36, 38.
19. [Thomas Carr in the] *Cobourg Star*, 22 February 1831, 53.
20. Need, *Six Years in the Bush*, 52 – 4.
21. *JHA*, 1833 – 1834, 11, 14.
22. *JHA*, 1835, 53.
23. Ibid.
24. Ibid., 55, 249.
25. Need, *Six Years in the Bush*, 109 – 10.
26. Peggy Blair, "Taken for 'Granted': Aboriginal Title and Public Fishing Rights in Upper Canada," *Ontario History* 92, 2 (Spring 2000): 31 – 55.
27. Elinor G. K. Melville, *A Plague of Sheep: Environmental Consequences of the Conquest of Mexico* (Cambridge and New York:

Cambridge University Press, 1994), esp. 116 – 50.

28. Peter Jones, 1831 letter, quoted in The Aborigines Protection Society, *Report on the Indians of Upper Canada* (London: William Ball, Arnold, and Company, 1839), 16.
29. Ibid.
30. An overview of this early system is provided by R. J. Surtees, "The Development of an Indian Reserve Policy in Canada," in *Historical Essays on Upper Canada*, ed. J. K. Johnson (Toronto: McClelland and Stewart Ltd., 1975), and by L.F.S. Upton, "The Origins of Canadian Indian Policy," *Journal of Canadian Studies* 8, (November 1973): 51 – 61.
31. *Appendix to the Fourth Volume of the Journals of the Legislative Assembly of the Province of Canada*, 1844 – 45, *Appendix, no.* 2, Appendix EEE, Section II, "Report on the Affairs of the Indians in Canada [20 Mar. 1845]", "[Evidence of Rev. W. Case, no. 9] The Mississagas [*sic*] of Alnwick."
32. [Thomas Carr in the] *Cobourg Star*, 15 February 1831, 46.
33. Ibid. The district native commercial fishery was perhaps short-lived, since by summer 1831 Cobourg men were using native-inspired seines to fish in Lake Ontario; see *Cobourg Star*, 28 June 1831, 197 – 8.
34. Traill, *The Backwoods of Canada*, 120 – 1.
35. Ibid., 174 – 5.
36. [Thomas Carr in the] *Cobourg Star*, 15 February 1831, 46.
37. Charles Richard Weld, *A Vacation Tour in the United States and Canada* (London: Longman, Brown, Green, and Longmans, 1855), 116.

Chapter 3

1. Catharine Parr Traill, *The Backwoods of Canada: Being Letters from the Wife of an Emigrant Officer, Illustrative of the Domestic Economy of British America* (London: Charles Knight, 1836; reprinted, Toronto: McClelland and Stewart, 1989), 48.
2. T. W. Freeman, *Ireland: Its Physical, Historical, Social and Economic Geography* (London: Methuen and Co.; New York: E. P. Dutton and Co., 1950), 28.
3. Ibid., 16 – 7, 26.
4. Ibid., 20, 28.
5. Ibid., 52, 57 – 9.
6. William J. Smyth, "Social, Economic and Landscape Transformations in County Cork from the Mid-Eighteenth to the Mid-Nineteenth Century," in *Cork History and Society: Interdisciplinary Essays on the History of an Irish County*, eds. Patrick O'Flanagan and Cornelius G. Buttimer (Dublin: Geography Publications, 1993), 673, 680, 683.
7. H.J.M. Johnston, *British Emigration Policy, 1815 – 1830: "Shovelling out Paupers"* (Oxford: Clarendon Press, 1972), 24; Aidan Clarke, with R. Dudley Edwards, "Pacification, Plantation, and the Catholic Question, 1603 – 23," in *A New History of Ireland*, vol. 3, *Early Modern Ireland*, 1534 – 1691, eds. T. W. Moody, F. X. Martin, and F. J. Byrne (Oxford: Clarendon Press, 1976), 193, 197.
8. James S. Donnelly, Jr., *The Land and the People of Nineteenth-Century Cork: The Rural Economy and the Land Question* (London: Routlege and Kegan Paul, 1975), 18 – 23.
9. Archives of Ontario (hereafter AO), F61, Peter Robinson Papers, [Peter Robinson] Report on Emigration, 4 May 1827 (MS 12/reel 3); Donnelly, *The Land and the People*, 18 – 28; K. H. Connell, "The Colonization of Waste Land in Ireland, 1780 – 1845," *Economic History Review*, 2d

ser., 3, 1 (1950): 44 – 71; Smyth, "Social, Economic and Landscape Transformations," 681 – 3.

10. Klaus E. Knorr, *British Colonial Theories, 1570 – 1850* (Toronto: University of Toronto Press, 1944), 269 – 315.
11. Gerald M. Craig, *Upper Canada: The Formative Years, 1784 – 1841* (McClelland and Stewart, 1963), 88, 126, 128; Peter L. Maltby and Monica Maltby, "A New Look at the Peter Robinson Emigration of 1823," *Ontario History* 55 (March 1963): 15 – 21; Johnston, *British Emigration Policy*, 1 – 56; Marianne McLean, *The People of Glengarry: Highlanders in Transition, 1745 – 1820* (Kingston and Montreal: McGill-Queen's University Press, 1991).
12. Wendy Cameron, "Selecting Peter Robinson's Irish Emigrants," *Historie sociale-Social History* 9 (May 1976), 29.
13. AO, F61, Kingston to Robinson, 19 December 1824 (MS 12/3).
14. For a description, see Stephen J. Pyne, *Fire in America: A Cultural History of Wildland and Rural Fire* (Princeton: Princeton University Press, 1982), 47.
15. Jonathan Bell and Mervyn Watson, *Irish Farming: Implements and Techniques, 1750 – 1900* (Edinburgh: John Donald, 1986), 205.
16. Michael Williams, *Americans and Their Forests: A Historical Geography* (Cambridge: Cambridge University Press, 1989), 60.
17. Samuel Strickland, *Twenty-Seven Years in Canada West; Or, The Experience of an Early Settler*, 2 Vols., ed. Agnes Strickland (London: R. Bentley, 1853; reprinted, Edmonton: M. G. Hurtig, Ltd., 1970), 1:91, 92 – 3.
18. *Upper Canada Herald* (Kingston) 29 September 1830, quoted in Peter A. Russell, "Forest into Farmland: Upper Canadian Clearing Rates, 1822 – 1839," *Agricultural History* 57 (July 1983), 338.
19. Traill, *The Backwoods of Canada*, 110.
20. Ibid., 130.
21. Robert Leslie Jones, *History of Agriculture in Ontario, 1613 – 1880*, (Toronto: University of Toronto Press, 1946; reprinted, Toronto: University of Toronto Press, 1977), 7 – 8.
22. Province of Canada, *Census Report of the Canadas, 1851 – 2*, vol. 2, no. 6, "Upper Canada Return of Agricultural Produce for 1851 – 2" (Québec: King's Printer, 1855), 40 – 3, 50 – 1.
23. Basil Hall, *Travels in North America in the Years 1827 and 1828*, 3 vols. (Edinburgh: Cadell and Co.; London: Simpkin and Marshall, 1829), 1:290, 288.
24. Williams, *Americans and Their Forests*, 61; William I. Roberts, III, "American Potash Manufacture before the American Revolution," American Philosophic Society, *Proceedings* 116 (October 1972): 383 – 95; Douglas McCalla, *Planting the Province: The Economic History of Upper Canada, 1784 – 1870* (Toronto: University of Toronto Press, 1993), passim; Alan Taylor, *William Cooper's Town: Power and Persuasion on the Frontier of the Early American Republic* (New York: Vintage, 1995), passim; Alan Taylor, "The Great Change Begins: Settling the Forest of Central New York," *New York History* 76 (July 1995): 265 – 90.
25. Thomas Need, *Six Years in the Bush; Or, Extracts from the Journal of a Settler in Upper Canada, 1832 – 1838* (London: Simpkin, Marshal and Co., 1838), 77 – 8.
26. Jones, *History of Agriculture in Ontario*, 32 – 3, 71 – 2; Howard T. Pammett, *Lilies and Shamrocks: A History of Emily Township, County of Victoria, Ontario, 1818 – 1973* (Lindsay, ON: John Deyell Co., 1974), 83; Harry Miller, "Potash from Wood Ashes: Frontier Technology in Canada and the United States," *Technology and Culture* 21 (April 1980): 187 – 208; Pyne, *Fire in America*, 128; Williams, *Americans and*

Their Forests, 61, 74 – 5; McCalla, *Planting the Province*, 49.
27. McCalla, *Planting the Province*, 22, 31, 39, 47.
28. Pammett, *Lilies and Shamrocks*, 22.
29. Williams, *Americans and Their Forests*, 74.
30. Catharine Parr Traill, *The Canadian Settler's Guide* (Toronto: Old Countryman Office, 1855; reprinted, Toronto: McClelland and Stewart, 1969), 167 – 73; Clifford B. Theberge and Elaine Theberge, *At the Edge of the Shield: A History of Smith Township, 1818 – 1980* (Peterborough, ON: Alger Press, Ltd. for the Smith Township Historical Committee, 1982), 41.
31. Isabel Skelton, *The Backwoodswoman: A Chronicle of Pioneer Home Life in Upper and Lower Canada* (Toronto: Ryerson Press, 1924), 228 – 31.
32. Ibid., 49.
33. *Cobourg Star*, 28 March 1832, 88.
34. *Peterborough Chronicle* 27 May 1845, 3; 3 Feb. 1846, 1.
35. Smyth, "Social, Economic and Landscape Transformations," 683.
36. Province of Canada, *Census Report of the Canadas*, 1851 – 2, vol. 2, no. 6, 8 – 11, 40 – 3, 48 – 51; Jones, *History of Agriculture in Ontario*, 86 – 9.
37. Jones, *History of Agriculture in Ontario*, 86.
38. Jones, *History of Agriculture in Ontario*, 101 – 5; Jean Murray Cole, *David Fife and Red Fife Wheat: The Story of a Few Grains of Wheat That Were Saved from a Cow* (Keene, ON: Lang Pioneer Village, 1992), 3 – 5.
39. Traill, *The Backwoods of Canada*, 153.
40. Ibid., 155.
41. William Cronon, *Changes in the Land: Indians, Colonists, and the Ecology of New England* (New York: Hill and Wang, 1983), 128.
42. David Dickson, "Butter Comes to Market: The Origins of Commercial Dairying in County Cork," in O'Flanagan and Buttimer, eds., *Cork History and Society*, 367; Smyth, "Social, Economic and Landscape Transformations," 655 – 6.
43. Province of Canada, *Census Report of the Canadas*, 1851 – 2, vol. 2, no. 6, 8 – 11, 40 – 3, 48 – 51. The breakdown in whole numbers is as follows: Douro 9,156 acres under cultivation to 3,476 under pasture; Smith 15,316 cultivated to 4,190 pasture; Ennismore 9,681 cultivated to 1,250 pasture; Emily 12,224 cultivated to 4,863 pasture; Ops 11,552 cultivated to 4,208 pasture; Asphodel 7,709 cultivated to 2,445 pasture; Otonabee 20,363 cultivated to 5,482 pasture.
44. John Mannion, *Irish Settlements in Eastern Canada: A Study of Cultural Transfer and Adaptation* (Toronto: University of Toronto Press, 1974), 40; Province of Canada, *Census Report of the Canadas*, 1851 – 2, vol. 2, no. 6, 10 – 1, 42 – 3, 50 – 1.
45. Traill, *The Backwoods of Canada*, 88.
46. Cronon, *Changes in the Lands*, 129.
47. Province of Canada, *Census Report of the Canadas*, 1851 – 2, vol. 2, no. 6, 10 – 1, 42 – 3, 50 – 1.
48. Archives of Ontario, MS 180, "Aggregate Census and Assessment Returns for Upper Canada, 1825 – 1849," "A General Account of the Rateable Property in the District of Newcastle for the Year Ending upon the First Monday in the Month of January, 1827."
49. Province of Canada, *Census Report of the Canadas*, 1851 – 2, vol. 2, no. 6, 10 – 1, 42 – 3, 50 – 1.
50. *Cobourg Star*, 6 and 13 May 1835, 2.
51. Ibid., 10 September 1834, 2.
52. Province of Canada, *Census Report of the Canadas*, 1851 – 2, vol. 2, no. 6, 10 – 1, 42 – 3, 50 – 1.
53. Levi Payne letter, Dummer Township, 13 October 1831, quoted in Edwin C. Guillet, ed., *The Valley of the Trent* (Toronto: Champlain Society, 1957), 71.

54. AO, MS 180,"A General Account of Rateable Property in the District of Newcastle for the Year Ending [1827]."
55. AO, MS 180,"Aggregate of Rateable Property in the Newcastle District for the Year 1836."
56. Province of Canada, *Census Report of the Canadas*, 1851 – 2, vol. 2, no. 7,"Upper Canada Return of Mills, Manufactories, and Etc., for 1851 – 2," 230 – 1.
57. National Archives of Canada (hereafter NA), Record Group 5, Series A1, Civil Secretary's Correspondence, Upper Canada Sundries, vol. 75, 40369 – 71, Petition of the Inhabitants of Emily to Sir Peregrine Maitland, [1825] (C-4616).
58. Traill, *The Backwoods of Canada*, 235.
59. Ibid.
60. NA, RG 5, C1, Canada West, Provincial Secretary's Correspondence, vol. 44, file 2464 1/2, 18612 – 30, Re: Burnham and Gilchrist operations at Stoney Lake, 1840 (C-13560).
61. "Directory of the United Counties of Peterborough and Victoria for 1858," quoted in Guillet, ed., *The Valley of the Trent*, 276.
62. See, for example, *Cobourg Star*, 5 April 1831, 102; 27 December 1831, 406. See also Thomas F. McIllwraith,"Transportation in the Landscape of Early Upper Canada," and Peter Ennals,"Cobourg and Port Hope: The Struggle for the 'Back Country,'" in *Perspectives on Landscape and Settlement in Nineteenth Century Ontario*, ed. J. David Wood (Ottawa: Carleton University Press, 1975).
63. *Cobourg Star*, 18 December 1833, 2; 2 July, 6 August, 10 October, 26 November 1834, 2; 15 July, 30 September, 11 November 1835, 2; 10 August, 26 October 1836, 2.
64. *Cobourg Star*, 9 November 1836, 2.
65. Ibid.
66. Donna Birdwell-Pheasant,"The Home 'Place': Center and Periphery in Irish House and Family Systems," in *House Life: Space, Place and Family in Europe*, ed. Donna Birdwell-Pheasant and Denise Lawrence-Zuniga (Oxford and New York: Berg, 1999), 105.
67. *Cobourg Star*, 9 October 1833, 2 – 3.
68. Alan G. Brunger,"Geographical Propinquity among Pre-Famine Catholic Irish Settlers in Upper Canada," *Journal of Historical Geography* 8, 3 (1982), esp. 277.
69. Ibid., 275.
70. Ibid., 270.
71. Mannion, *Irish Settlements*, 74, 79.
72. Hall, *Travels in North America*, 1: 298 – 9. Canal work was an attractive option for some Emily and Ennismore township men; see AO, RG1, Crown Lands Department Correspondence, series A-I-6, vol. 8, env. 2, 7315 – 8, Alexander McDonell to Robinson, 1 July 1828 (MS 563/8).
73. Hall, *Travels in North America*, 1: 298 – 9.
74. Ibid., 1: 299.
75. NA, RG5, A1, vol. 75, 40372 – 5, Irish Emigrants of Asphodel to Earl of Bathurst, [1825] (C-4616). See also NA, RG5, A1, vol. 80, 43439 – 42, Inhabitants of Douro to Earl of Bathurst, 12 December 1826 (C-6862).

Chapter 4

1. Raymond Williams, *Problems in Materialism and Culture: Selected Essays* (London: Verso, 1980), 82.
2. James T. Angus, *A Respectable Ditch: A History of the Trent-Severn Waterway, 1833 – 1920* (Kingston and Montreal: McGill-Queen's University Press, 1988).
3. John Langton, *Early Days in Upper Canada, Letters of John Langton from the Backwoods of Upper Canada and the Audit Office of the Province of Canada*, ed. W. A. Langton

(Toronto: Macmillan, 1926), 47 – 8. Purdy's Mills began to be referred to as Lindsay in 1836.

4. L. J. Chapman and D. F. Putnam, *The Physiography of Southern Ontario*, 2d ed. (Toronto: University of Toronto Press, 1966), 296 – 9.
5. Moti Tahiliani, ed., *"Bless These Walls:" Lindsay's Heritage* (Lindsay, ON: Blewet Printing, 1982), 21.
6. Ontario Department of Lands and Forests (hereafter ODLF), Land and Survey Papers, *Report and Field Notes, no.* 571, Duncan McDonell, "Field Notes of the Township of Ops, no. 1, 1825."
7. National Archives of Canada (hereafter NA), Civil Secretary Correspondence, RG5, series A1, vol. 94, 52790 – 3, Memorial of Duncan McDonell, 24 June 1829 (C-6868); Archives of Ontario (hereafter AO), Crown Lands Department, RG1, Subject Files, series A-I-7, Settlement Files, vol. 16, env. 15, 10685 – 701, Ops Township, 1826 (MS 892, reel 11). My emphasis.
8. Frederick L. Dunn, "Malaria," in *The Cambridge World History of Human Disease*, ed. Kenneth F. Kiple (Cambridge: Cambridge University Press, 1993), 855 – 62; David P. Adams, "Malaria, Labor, and Population Distribution in Costa Rica: A Biohistorical Perspective," *Journal of Interdisciplinary History* 27 (Summer 1996): 75 – 7; Herbert M. Gilles and David A. Warrell, eds., *Bruce-Chwatt's Essential Malariology*, 3d ed. (London: Edward Arnold, 1993), 20 – 7, 44 – 5, 96, 102, 111, 113.
9. A. Murray Fallis, "Malaria in the 18th and 19th Centuries in Ontario," *Bulletin canadien d'historie de la medecine/Canadian Bulletin of Medical History* 1 (Hiver/Winter 1984): 25 – 38; Robert Legget, *Rideau Waterway* (Toronto: University of Toronto Press, 1955), 51 – 2, 54, 116, 127, 185, 210; William N. T. Wylie, "Poverty, Distress, and Disease: Labour and the Construction of the Rideau Canal, 1826 – 32," *Labour/Le Travail* 11 (Spring 1983): 7 – 29; R. Cole Harris, Pauline Roulson, and Chris De Freitas, "The Settlement of Mono Township," *Canadian Geographer* 19, 1 (1975): 10; NA, Canada West, RG5, Provincial Secretary Correspondence, series C1, vol. 190, file 14936, 29401 – 4, Dominic Daley to [Secretary of Public Works] W. B. Robinson, 17 September 1846 (H-2369); vol. 193, file 15335, 30764 – 8, Daley to [Secretary of Public Works] Thomas A. Begley, 16 February 1847 (H-2370); Anne Langton, *A Gentlewoman in Upper Canada: The Journals of Anne Langton*, ed. H. H. Langton (Toronto: Clarke, Irwin and Co. Ltd., 1950; Toronto: Irwin Publishing, 1964), 135, 192; Norman D. Levine introduction to *Malaria in the Interior Valley of North America* in Daniel Drake, *A Systematic Treatise, Historical, Etiological, and Practical, on the Principal Diseases of the Interior Valley of North America, as They Appear in the Caucasian, African, Indian, and Esquimaux Varieties of the Population* (Cincinnati: Winthrop B. Smith and Co.; Philadelphia: Grigg, Elliot and Co.; New York: Mason and Law, 1850; reprinted, Urbana: University of Illinois Press, 1964), xviii; Erwin H. Ackerknecht, *Malaria in the Upper Mississippi Valley, 1760 – 1900* (Baltimore: The Johns Hopkins University Press, 1945); Robert T. Boyd, "Another Look at the 'Fever and Ague' of Western Oregon," *Ethnohistory* 22 (Spring 1975): 135 – 54.
10. Langton, *A Gentlewoman*, 171 – 2.
11. AO, RG1, Correspondence, series A-I-6, vol. 8, env. 5, 7650 – 1, Peter Robinson to Wellesly Rickey, 5 June 1829 (MS 563/8).
12. AO, RG1, A-I-6, vol. 8, env. 5, 7666 – 9, Alexander McDonell to Robinson, 26 June 1829 (MS 563/8).

13. NA, RG5, Civil Secretary Correspondence, series A1, vol. 94, 52794 – 7, McDonell to Robinson, 22 June 1829 (C-6868); AO, RG1, A-I-6, vol. 8, env. 5, 7692 – 5, McDonell to Robinson, 9 July 1829; 7706 – 9, McDonell to Robinson, 21 July 1829; 7713 – 5, McDonell to Robinson, 24 July 1829 (MS 563/8).
14. AO, RG1, A-I-6, vol. 8, env. 5, 7713 – 5, McDonell to Robinson, 24 July 1829 (MS 563/8).
15. AO, RG1, A-I-6, vol. 9, env. 3, 8189 – 92, McDonell to Robinson, 1 July 1830; 8197 –8200, McDonell to Robinson, 2 July 1830 (MS 563/8).
16. AO, RG1, A-I-6, vol. 8, env. 5, 7734 – 6, McDonell to Robinson, 16 August 1829 (MS 563/8).
17. AO, RG1, A-I-6, vol. 8, env. 5, 7751 – 3, McDonell to Robinson, 21 August 1829; 7760 – 2, McDonell to Robinson, 24 August 1829; env. 6, Thomas Baines to Robinson, 16 September 1829; McDonell to Robinson, 17 October 1829 (MS 563/8).
18. NA, RG5, A1, vol. 96, 53981 – 8, "List of Emigrants located in the Township of Ops by Mr. A. McDonell up to the 15th of October 1829," 29 October 1829 (C-6868).
19. AO, RG1, A-I-6, vol. 8, env. 6, 7815 – 8, McDonell to Robinson, 21 October 1829 (MS 563/8).
20. AO, RG1, A-I-6, vol. 8, env. 6, 7850 – 4, Robinson to McDonell, 26 November 1829; 7870–2, McDonell to Robinson, 11 December 1829; 7887 – 8, McDonell to Robinson, 20 December 1829 (MS 563/8); AO, RG1, Lands Branch, Applications, series C-I-1, vol. 39, "Petition of John Malony and Others to Peter Robinson, 1829," (MS 691/51).
21. AO, RG1, A-I-6, vol. 9, env. 1, 7911 – 2, McDonell to Robinson, 15 January 1830 (MS 563/8); NA, RG5, A1, vol. 98, 54831 – 2, Robert Blaylock to Sir John Colborne, 9 January 1830 (C-6869).
22. AO, RG1, A-I-6, vol. 9, env. 3, 8273 – 6, McDonell to Robinson, 25 August 1830; AO, RG1, A-I-6, vol. 9, env. 4, 8292 – 5, McDonell to Robinson, 2 September 1830 (MS 563/9).
23. Morton Horwitz, *The Transformation of American Law, 1780 – 1860* (Cambridge, MA: Harvard University Press, 1977); Gary Kulik, "Dams, Fish, and Farmers: Defense of Public Rights in Eighteenth-Century Rhode Island," in *The Countryside in the Age of Capitalist Transformation*, ed. Steven Hahn and Jonathan Prude (Chapel Hill: University of North Carolina Press, 1985), 25 – 50. Brian Donahue, "'Damned at Both Ends and Cursed in the Middle': The 'Flowage' of the Concord River Meadows, 1798 – 1862," *Environmental Review* 13 (Fall/Winter 1989): 47 – 67; Theodore Steinberg, *Nature Incorporated: Industrialization and the Waters of New England* (Cambridge and New York: Cambridge University Press, 1991); Harry L. Watson, "'The Common Rights of Mankind': Subsistence, Shad, and Commerce in the Early Republican South," *Journal of American History* 83 (June 1996): 13 – 43.
24. Thomas Need, *Six Years in the Bush; or, Extracts from the Journal of a Settler in Upper Canada, 1832 – 1838* (London: Simpkin, Marshall and Co., 1838), 96; Arthur R. M. Lower, *Settlement and the Forest Frontier in Eastern Canada*, vol. 9 of *Canadian Frontiers of Settlement*, ed. W. A. Mackintosh and W.L.G. Joerg (Toronto: Macmillian, 1936), 41; Louis C. Hunter, *A History of Industrial Power in the United States, 1780 – 1930*, vol. 1 of *Waterpower in the Century of the Steam Engine* (Charlottesville: University Press of Virginia, 1979), 3, 28; see also Felicity L. Leung, *Grist and Flour Mills in Ontario: From Millstones to Rollers, 1780s – 1880s*, History and Archaeology Series no. 53 (Ottawa: Parks Canada, 1981).

25. Elwood Jones,"Purdy, William," in *Dictionary of Canadian Biography*, vol. 7 (Toronto: University of Toronto Press, 1988): 712 – 3; NA, Upper Canada Land Petitions, RG1, series L3, vol. 418,'P' Bundle Miscellaneous, 1775 – 95, nos. 182, 182a (C-2737); NA, RG1, L3, vol. 402,'P' Bundle 8, 1806 – 8, no. 64 (C-2490); Death Notice, *Christian Guardian*, 7 April, 12 May 1847; George Elmore Reaman, *A History of Vaughan Township: Two Centuries of Life in the Township* (Toronto: C. H. Snider, 1971), 54 – 5.
26. NA, RG5, A1, vol. 98, 55323 – 5, William Purdy to Robinson, 16 February 1830 (C-6869).
27. NA, RG5, A1, vol. 98, 55319 – 22, Robinson to [Civil Secretary] Zachariah Mudge, 17 February 1830 (C-6869).
28. NA, RG5, A1, vol. 98, 55219 – 21, Robinson to Mudge, 7 February 1830 (C-6869); AO, RG1, A-I-6, vol. 9, env. 4, 8288 – 91, Purdy to Robinson, 1 September 1830 (MS 563/9).
29. AO, RG1, A-I-6, vol. 9, env. 4, 8336 – 8, McDonell to Robinson, 4 October 1830; 8415 – 7, McDonell to Robinson, 26 November 1830; vol. 9, env. 5, 8434 – 7, McDonell to Robinson, 2 December 1830; 8452 – 5, McDonell to Robinson, 15 December 1830 (MS 563/9); vol. 11, env. 7, 10231 – 2, Robinson to [Civil Secretary] William Rowan, 21 May 1833 (MS 563/10).
30. NA, RG5, A1, vol. 106, 60463 – 6, Purdy to Sir John Colborne, 14 April. 1831 (C-6072).
31. NA, RG5, A1, vol. 106, 60467 – 8, Robinson to Mudge, 21 April 1831 (C-6072); AO, RG1, A-I-6, vol. 9, env. 7, 8659 – 62, McDonell to Robinson, 5 May 1831.
32. AO, RG1, A-I-6, vol. 10, env. 3, 9016 – 8, McDonell to Robinson, 26 December 1831 (MS 563/9).
33. AO, RG1, Township Papers, series C-IV, box 369, 001194–6, env. 6, McDonell to Robinson, 12 December 1829 (MS 658/356).
34. AO, Cobourg, Newcastle District Court of General Quarter Sessions of the Peace, RG22, Road and Bridge Records, series 35, box 5, 1832, no. 269.
35. AO, RG22-35, box 6, no. 299, Report of H. Ewing, 14 January 1834; box 7, no. 380, Report of J. Huston, 10 October 1836.
36. Gerald M. Craig, *Upper Canada: The Formative Years, 1784 – 1841* (Toronto: McClelland and Stewart, 1963), 150 – 60.
37. Angus, *A Respectable Ditch*, 3 – 4.
38. Ibid., 5 – 11. Rear township sentiment is most evident in NA, RG5, A1, vol. 171, 93204 – 11, Robert Jameson to John Joseph, 1 October 1836 (C-6891), and *Cobourg Star*, 26 October, 9 November 1836, 2. Other salient references include *Cobourg Star*, 30 October 1833, 2; 5 February, 2 July, 6 August, 10 October, 26 November 1834, 2; 15 July, 30 September, 11 November 1835, 2; 16 March, 7, 21 December 1836, 2; 8, 15 February 1837, 2.
39. "[Commissioners for the Improvement of the Navigation of the Inland Waters of the District of Newcastle] Petition Forwarded to His Excellency," *Journal of the House of Assembly of Upper Canada* (hereafter *JHA*), 1833 – 1834, 204 – 5.
40. NA, RG5, A1, vol. 135, 74271 – 7, James Gray Bethune to Rowan, 23 November 1833 (C-6880).
41. NA, RG5, A1, vol. 155, 85019 – 20, Petition of Inhabitants of the Township of Ops, 18 July 1835 (C-6886).
42. NA, RG1, L3, vol. 408, Upper Canada Land Petitions 'P' Bundle 18, 1833 – 5, nos. 107 – 107a, Purdy to Sir John Colborne, 5 October 1833 (C-2731A).
43. NA, RG1, L3, vol. 408, no. 107c, Letter of Alexander McDonell, 5 October 1833 (C-2731A).
44. NA, RG1, L3, vol. 408, no. 107b, Letter of Peter Robinson, 6 May 1834 (C-2731A).

45. NA, RG1, L3, vol. 408, no. 107b, Letter of John Beike [Clerk of General Court], 9 May 1834 (C-2731A).
46. Peter George and Philip Sworden, "The Courts and the Development of Trade in Upper Canada, 1830 – 1860," *Business History Review* 60 (Summer 1986), 274; Peter George and Philip Sworden, "John Beverly Robinson and the Commercial Empire of the St. Lawrence," *Research in Economic History* 11 (1988), 231 – 2.
47. *British Whig* 4 November 1834, 3; *Chronicle and Gazette* 27 December 1834, 3; Upper Canada, *Journals of the Legislative Council of Upper Canada*(hereafter *JLC*), 1835, 20 January 1835, 32; 22 January 1835, 40.
48. *JHA*, 1835, 2 March 1835, 195.
49. *JHA*, 1835, 9 March 1835, 218.
50. *JHA*, 1835, 15 April 1835, 399 – 400; *Appendix to Journal of the House of Assembly of Upper Canada*, 1835, vol. 2, no. 99, "Report of the Select Committee of the Petitions of William Purdy and Robert Jameson," 15 April 1835.
51. Wendy Cameron, "Nicol Hugh Baird and the Construction of the Trent-Severn Waterway," in *Canadian Papers in Rural History*, Vol. 7, ed. Donald H. Akenson (Gananoque, ON: Langdale Press, 1990), 257 – 72.
52. AO, Baird Family Papers, F645, Trent Canal Miscellaneous Documents, Scugog River, 1835 – 43, series A-6-b (ii), box 9, env. 11, "Report on the Overflowing of the Scugog River and Lake, and on the Effect the Removal of the Dam at Purdy's Mill Would Have Upon the Navigation of the Scugog River and Lake [10 – 18 Oct. 1835]" (MS 393/5). This report was reprinted in Upper Canada, *Appendix to the Journal of the House of Assembly*, 1836 (hereafter *Appendix*), vol. 1, no. 13, 1. To avoid adding errors of my own I will quote from the printed source.
53. *Appendix*, 2, 5.
54. Ibid., 3.
55. Ibid., 5; NA, Department of Railways and Canals, RG43, Canals Branch Correspondence, vol. 223, file 18344, part 2, no. 17, Baird to Begley, 15 August 1842.
56. *Appendix*, 2.
57. Ibid., 6.
58. Ibid., 3.
59. Ibid., 4, 7.
60. Tahiliani, ed., *"Bless These Walls"*, 2.
61. Angus, *A Respectable Ditch*, 38.
62. *Appendix*, 7 – 8.
63. *Cobourg Star*, 7 December 1836, 2.
64. Trent University Archives, John Huston Fonds, 71 – 006, series A, box 2, folder 3, no. 271, John Logie to John Huston, 9 September 1836.
65. *JHA*, 1836 – 1837, "Petition of John Logie and 67 Other Inhabitants of Ops Township," 2 December 1836, 129, 141; 7 December 1836, 33; "Petition of William Purdy and 135 Others of Ops Township," 12 December 1836, 161.
66. Langton, *Early Days in Upper Canada*, 47, 107, 147.
67. NA, RG5, A1, vol. 99, 55599-600b, List of Magistrates for the Newcastle District, 13 July 1830 (C-6869); vol. 104, 58935 – 7, Thomas A. Stewart to Mudge, 18 December 1830 (C-6871); vol. 126, 69525 – 35, Petition of Ops Inhabitants, 17 February 1833 (C-6877); vol. 145, 79407 – 19, Recommendations for Newcastle District Magistrates, 15 September 1834 (C-6883).
68. *Cobourg Star*, 5 April 1831, 102; 1 February 1832, 22; 12 March 1834, 3.
69. AO, RG22, Petitions 1835 – 7, series 37, box 1, "Newcastle District (Ops Township), Large Number of Ratepayers [Petition] January 1837."
70. Ibid.
71. AO, RG22-37, box 1, "Petition of Freeholders and Other Inhabitants of the Township of Ops, January 1837."

72. AO, RG1, A-I-7, Mill Sites, vol. 10, env. 7, no. 15, 05762, Petition of William Purdy [to Sir Francis Bond Head], 31 January 1837 (MS 892/6).
73. AO, RG1, A-I-7, vol. 10, env. 7, no. 15, 05764, Letter of Alexander McDonell, 2 February 1837 (MS 892/6).
74. Ibid.
75. AO, F645, A-6-a, box 4, env. 2, William Boulton to Purdy, 18 November 1837 (MS 393/2).
76. Quentin Brown, "Swinging with the Governors: Newcastle District Elections, 1836 and 1841," *Ontario History* 86 (December 1994): 319 – 36.
77. NA, RG5, C1, vol. 9, file 1155, 4887 – 9, McDonell to Joseph, 6 December 1837 (C-13552).
78. Watson Kirkconnell, *County of Victoria Centennial History*, rev. ed. (Lindsay, ON: John Deyell, 1967), 103 – 5.
79. *Cobourg Star*, 13 December. 1837, 1.
80. Catharine Parr Traill, "The Mackenzie Rebellion," in *Forest and Other Gleanings: The Fugitive Writings of Catharine Parr Traill*, ed. Michael A. Peterman and Carl Ballstadt (Ottawa: University of Ottawa Press, 1994), 120.
81. Kirkconnell, *County of Victoria*, 104 – 5.
82. *Cobourg Star*, 20 December 1837, 3; AO, Correctional Services, RG20, series F-9, vol. 1, Cobourg Jail Register, 1834 – 48. A subsequent government report noted that Purdy was jailed for "seditious language" and not treason, see NA, RG5, A1, vol. 201, 111358 – 61, [Newcastle District Sheriff] Charles S. Ruttan to [Civil Secretary] John Macaulay, 9 August 1838 (C-6901).
83. *Christian Guardian* 7 April 1847, 2; 12 May 1847, 3.
84. *Cobourg Star*, 2 May 1838, 3.
85. "Report of the Commissioners for the Improvement of the Navigation of the Inland Waters of the District of Newcastle [26 Jan. 1839]," in *Appendix to the Journal of the House of Assembly of Upper Canada*, 1839, vol. 1, part 1, 156.
86. Hamilton H. Killaly, "Report of Engineer on the Survey of Country Lying Between Lake Ontario and Lakes Scugog and Simcoe [26 February 1838]," in *Appendix to the Journal of the House of Assembly of Upper Canada*, 1837 – 1838, 386 – 7; *Cobourg Star*, 8 August 1838, 3.
87. AO, F645, A-6-a, box 4, env. 3, [canal inspector] Thomas McNeil to Baird, 8 February 1838; box 4, env. 4, McNeil to Baird, 12 October 1838; McNeil to Baird, 30 October 1838 (MS 393/2).
88. AO, F645, A-6-a, box 4, env. 7, Homer Hecox to Baird, 2 October 1839 (MS 393/3).
89. AO, F645, A-6-a, box 4, env. 3, McNeil to Baird, 8 February 1838; box 4, env. 4, McNeil to Baird, 30 October 1838; box 4, env. 7, Baird to Commissioners for the Improvement of the Inland Waters of the Newcastle District, 10 September 1839 (MS 393/2).
90. *Cobourg Star*, 20 November 1839, 2; NA, RG43, vol. 222, file 18344, part 1, Baird to Commissioners for the Improvement of the Inland Waters of the District of Newcastle, 22 November 1839; "Report of Commissioners on the Improvement of the Navigation of the Inland Waters of the District of Newcastle," in *Appendix to the Journal of the House of Assembly*, 1840, vol. 1, 266.
91. AO, F645, A-6-a, box 4, env. 8, [Lindsay clerk] Thomas Clark to Baird, 13 April 1840; Clark to Baird, 9 June 1840 (MS 393/3); NA, RG5, C1, vol. 40, file 2064, 17234 – 7, Clark to Sir George Arthur, November 1840; McDonell to [Secretary] J.B. Harrison, 18 November 1840 (C-13559).
92. *Cobourg Star*, 2 June 1841, 2.
93. Ibid., 30 June, 28 July, 18 August 1841, 3. I take this as a reference to Ops. During

spring 1841, a Langton family servant became ill "after going for a day or two into Ops, which is a dreadfully unhealthy township"; see Langton, *A Gentlewoman*, 159.

94. NA, RG5, C1, vol. 89, file 3846, 34661 – 2, Frederick Widder to Harrison, 23 May 1842 (C-13570).
95. Ibid.
96. Ibid.
97. Ibid.
98. NA, RG5, C1, vol. 79, file 2525, 30434, Logie to [Secretary] James Hopkirk, 4 January 1842 (C-13568).
99. NA, RG5, C1, vol. 89, file 3846, 34643, Widder to Harrison, 27 April 1842 (C-13570).
100. NA, RG5, C1, vol. 89, file 2798, 34655, "[John Langton][Authorized Copy of the Adopted] Report by the Municipal Council of the Colborne District," 10 February 1842 (C-13570); *Cobourg Star*, 16 February 1842, 2.
101. Ibid.
102. Ibid.
103. NA, RG5, C1, vol. 83, file 2994, 32342 – 3, Harrison to John D'Arcus, 1 March 1842 (C-13569).
104. Jones, "Purdy, William," 713; [Hamilton H. Killaly] "Report of the Board of Works, 1844," Province of Canada, *Journal of the Legislative Assembly*, 1844 – 1845, *Appendix A. A.*, 1845.
105. Carol Wilton, *Popular Politics and Political Culture in Upper Canada*, 1800 – 1850 (Montreal: McGill-Queen's University Press, 2000). See also J. K. Johnson, "'Claims of Equity and Justice': Petitions and Petitioners in Upper Canada, 1815 – 1840," *Historie sociale/Social History* 28 (May 1995): 219 – 40; F. H. Armstrong, "The York Riots of March 23, 1832," *Ontario History* 55 (June 1963): 61 – 72; John Weaver, "Crime, Public Order, and Repression: The Gore District in Upheaval, 1832 – 1851," *Ontario History* 78 (September 1986): 175 – 207; Paul Romney, "From the Types Riot to the Rebellion: Elite Ideology, Anti-Legal Sentiment, Political Violence, and the Rule of Law in Upper Canada," *Ontario History* 79 (June 1987): 113 – 44; Leo A. Johnson, "The Gore District 'Outrages,' 1826 – 1829: A Case Study in Violence, Justice, and Political Propaganda," *Ontario History* 83, 2 (June 1991): 107 – 26; Susan Lewthwaite, "Violence, Law, and Community in Rural Upper Canada," in *Crime and Criminal Justice in Canadian History*, vol. 5 of *Essays in the History of Canadian Law*, ed. Jim Phillips, Tima Loo, and Susan Lewthwaite (Toronto: University of Toronto Press and the Osgoode Society for Canadian Legal History, 1994), 353 – 86; Edmund S. Morgan, *Inventing the People: The Rise of Popular Sovereignty in England and America* (New York: W. W. Norton, 1988), 209 – 33.
106. An exploration into this literature is provided by Scott W. See, "Nineteenth-Century Collective Violence: Toward a North American Context," *Labour/Le Travail* 39 (Spring 1997): 13 – 38. Notable studies include Michael S. Cross, "Stony Monday, 1849: The Rebellion Losses Riots in Bytown," *Ontario History* 63 (September 1971): 177 – 90; Michael S. Cross, "The Shiner's War: Social Violence in the Ottawa Valley in the 1830s," *Canadian Historical Review* 54 (March 1973): 1 – 26; Michael S. Cross, "'The Laws are Like Cobwebs': Popular Resistance to Authority in Mid-Nineteenth Century North America," in *Law in a Colonial Society: The Nova Scotia Experience*, ed. Peter B. Waite, Sandra Oxner, and Thomas Barnes (Toronto: Carswell, 1984), 103 – 23; Terence Crowley, "'Thunder Gusts': Popular Disturbances in Early French Canada," Canadian Historical Association *Historical Papers* (1979): 11 – 32; Ruth

Bleasdale, "Class Conflict on the Canals of Upper Canada in the 1840s," *Labour/Le Travailleur* 7 (1981): 9 – 89; Gregory S. Kealey, "Orangemen and the Corporation," in *Forging a Consensus: Historical Essays on Toronto*, ed. Victor L. Russell (Toronto: University of Toronto Press, 1984), 41 – 86; Stephen Kenny, "'Cahots' and Catcalls: An Episode of Popular Resistance in Lower Canada at the Outset of the Union," *Canadian Historical Review* 65 (June 1984): 184 – 208; Bryan D. Palmer, *Working-Class Experience: Rethinking the History of Canadian Labour, 1800 – 1991*, 2d ed. (Toronto: McClelland and Stewart, 1992), 35 – 116; Allan Greer, *The Patriots and the People: The Rebellion of 1837 in Rural Lower Canada* (Toronto: University of Toronto Press, 1993); Scott W. See, *Riots in New Brunswick: Orange Nativism and Social Violence in the 1840s* (Toronto: University of Toronto Press, 1993).

107. See Bryan D. Palmer, "Discordant Music: Charivaris and Whitecapping in Nineteenth-Century North America," *Labour/Le Travailleur* 3 (1978): 5 – 62.
108. NA, RG43, vol. 223, file 18344, part 2, no. 17, Baird to Begley, 15 August 1842.
109. Ibid.
110. NA, RG5, A1, vol. 209, 115307 – 10, Adam Meyers to Macauley, 10 November 1838 (C-6903); v. 210, 115761 – 3, E. Sanford to Arthur, 26 November 1838; 115791 – 3, J. Ham to Macauley, 26 November 1838; v. 224, 122931 – 50, Baron de Rottenberg to Arthur, 9 July 1839 (C-6907); v. 235, 128814 – 58, Rottenberg to Harrison, 14 December 1839 (C-6910); v. 241, 132104 – 14, Rottenberg letter, 24 March 1840 (C-6912); NA, RG5, C1, vol. 18, file 2165, 9538 – 41, Meyers to Macauley, 7 June 1839 (C-13555).
111. NA, RG5, A1, vol. 211, 116288 – 91, James Wallis to Charles Berczy, 12 December 1838 (C-6904).
112. William Loe Smith, *The Pioneers of Old Ontario* (Toronto: G. N. Morang, 1923), 228. See also F. G. Weir, *Scugog and its Environs* (Port Perry, ON: Star Print, 1927), 13 – 4; Samuel Farmer, *On the Shores of Scugog* (Port Perry, ON: Port Perry Star, 1934), 197 – 8.
113. ODLF, *Report and Field Notes, no.* 572, John Ryan, "Diary and Field Notes of a Survey of Some Lands Overflowed by the Scugog," 14 March 1849; AO, RG1, C-IV, box 369, env. 4, 1038, James Dennehy letter, 8 March 1858 (MS 658/356); AO, RG1, C-I-1, vol. 39, no. 585, Ops Inhabitants to [Governor General] Charles Metcalfe, 1 July 1845; no. 782, Patrick Hoey and Others to the Governor General, 22 May 1851; Bryan Hoey to Commissioner of Crown Lands, 15 June 1851 (MS 691/51); AO, RG1, A-I-6, vol. 31, env. 2, 26890–1, John Pyne to Commissioner of Crown Lands, 15 June 1854 (MS 563/28).

Chapter 5

1. Laurence Fallis, Jr., "The Idea of Progress in the Province of Canada: A Study in the History of Ideas," in *The Shield of Achilles: Aspects of Canada in the Victorian Age*, ed. W. L. Morton (Toronto: McClelland and Stewart, 1968); Carl Berger, *The Sense of Power: Studies in the Ideas of Canadian Imperialism* (Toronto: University of Toronto Press, 1970); Doug Owram, *Promise of Eden: The Canadian Expansionist Movement and the Idea of the West, 1856 – 1900* (Toronto: University of Toronto Press, 1980); Suzanne Zeller, *Inventing Canada: Early Victorian Science and the Idea of Transcontinental Nation* (Toronto: University of Toronto Press, 1987). Cf. J. I. Little, *Nationalism, Capitalism and*

Colonization in Nineteenth-Century Quebec: The Upper St. Francis District (Kingston and Montreal: McGill-Queen's University Press, 1989).

2. A.R.M. Lower, "Settlement and the Forest Frontier in Eastern Canada," in *Canadian Frontiers of Settlement*, vol. 9, ed. W. A. Mackintosh and W.L.G. Joerg (Toronto: University of Toronto Press, 1936), esp. 28 – 47; Philip L. White, *Beekmantown, NY: Forest Frontier to Farm Community* (Austin and London: University of Texas Press, 1979), 53 – 70; Graeme Wynn, *Timber Colony: A Historical Geography of Early Nineteenth Century New Brunswick* (Toronto: University of Toronto Press, 1981), 54 – 86; Richard W. Judd, *Aroostook: A Century of Logging in Northern Maine* (Orono: University of Maine Press, 1989), esp. 81 – 101; Beatrice Craig, "Agriculture and the Lumberman's Frontier in the Upper St. John Valley, 1800 – 70," *Journal of Forest History* 32 (July 1988): 125 – 37.
3. L. J. Chapman and D. F. Putnam, *The Physiography of Southern Ontario*, 2d ed. (Toronto: University of Toronto Press, for the Ontario Research Foundation, 1966), 313 – 6.
4. Miles Ecclestone, "The Physical Landscape of Peterborough and the Kawarthas," in *Peterborough and the Kawarthas*, ed. Peter Adams and Colin Taylor (Peterborough, ON: Heritage Publications, 1985), 25.
5. Cited by H. R. Cummings, *Early Days in Haliburton* (Toronto: Ryerson University Press, for the Ontario Department of Lands and Forests, 1962), 6. Cf. Richard B. Anderson, "Making Wilderness Smile: Professional Resource Evaluation in Victorian Canada," (Ph.D. diss., York University, 1992).
6. Lillian F. Gates, *Land Policies of Upper Canada* (Toronto: University of Toronto Press, 1968), 287 – 8. See also Richard S. Lambert with Paul Pross, *Renewing Nature's Wealth: A Centennial History of the Public Management of Lands, Forests and Wildlife in Ontario, 1763 – 1967* (Toronto: Hunter Rose Co., for the Ontario Department of Lands and Forests, 1967), 88.
7. Robert L. Jones, *History of Agriculture in Ontario, 1613 – 1880* (Toronto: University of Toronto Press, 1946; Toronto: University of Toronto Press, 1977), 291; J. Howard Richards, "Lands and Policies: Attitudes and Controls in the Alienation of Lands in Ontario During the First Century of Settlement," *Ontario History* 50 (Autumn 1958): 193 – 209; Florence B. Murray, "Agricultural Settlement on the Canadian Shield: Ottawa River to Georgian Bay," in *Profiles of a Province: Studies in the History of Ontario*, ed. Edith Firth (Toronto: Ontario Historical Society, 1967); Gates, *Land Policies*, 287; Keith A. Parker, "Colonization Roads and Commercial Policy," *Ontario History* 67 (March 1975): 31 – 8; Graeme Wynn, "Notes on Society and Environment in Old Ontario," *Journal of Social History* 13 (Fall 1979): 49 – 65; Helen E. Parson, "The Colonization of the Southern Canadian Shield in Ontario: The Hastings Road," *Ontario History* 79 (September 1987): 265 – 73; and John Andrew Leonard Macdonald, "A Region's Roots: Settlement and Land Alienation Along the Bobcaygeon Road, 1858 – 1900," (master's thesis, York University, 1994). Macdonald confines his study to Somerville and Stanhope townships.
8. Paul W. Gates, "Official Encouragement to Immigration by the Province of Canada," *Canadian Historical Review* 15 (March 1934): 24.
9. Canada, *Journals of the Legislative Assembly*, 1857, Appendix 54.
10. George W. Spragge, "Colonization Roads in Canada West, 1850 – 1867," *Ontario*

History Papers and Records 49 (Winter 1957): 10.

11. *Weekly Despatch* (Peterborough) 20 March 1851, 3. Cf. 6 March 1851, 2.
12. See *Peterborough Review* 13 January 1854, 2; 20 January 1854, 2; 27 January 1854, 2; 3 February 1854, 2; 17 March 1854, 2; 31 March 1854, 3; 13 April 1854, 2; 20 October 1854, 2; 10 November 1854, 2; 13 June 1856, 2; 20 June 1856, 2; 12 December 1856, 2; 27 February 1857, 2.
13. Thomas Need, *Six Years in the Bush; or, Extracts from the Journal of a Settler in Upper Canada, 1832–1838* (London: Simpkin, Marshall and Company, 1838), iii, 37–8.
14. Ibid., 96.
15. Archives of Ontario (hereafter AO), MU2186, Accession no. 6320, F-528, Thomas Need Papers, 1832–1883, E-1 Diary (1839).
16. *Census Report of the Canadas*, 1851–2, vol. 2, no. 7, "Upper Canada Returns of Mills, Manufactories, etc.," County of Peterborough, County of Victoria (Québec: King's Printer, 1855), 231, 243. In 1851, Peterborough County contained the townships of Belmont, Burleigh, Douro, Dummer, Harvey, Methuen, Smith, Monaghan, Asphodel, Ennismore, Otonabee, and the town of Peterborough. The townships of Mariposa, Ops, Emily, Eldon, Fenelon, Bexley, Verulam, and Somerville constituted Victoria County.
17. Charles Richard Weld, *A Vacation Tour in the United States and Canada* (London: Longman, Brown, Green, and Longmans, 1855), 99.
18. Thomas White, *An Exhibit of the Progress, Position and Resources of the County of Peterboro', Canada West, Based Upon the Census of 1861; Together with a Statement of the Trade of the Town of Peterborough* (Peterborough, ON: T & R White Printers, 1862), 17, 22.
19. Thomas W. Poole, *A Sketch of the Early Settlement and Subsequent Progress of the Town of Peterborough, and of Each Township in the County of Peterborough* (Peterborough, ON: Peterborough Review, 1867; reprinted, Peterborough, ON: Peterborough Printing Co., 1941), 92.
20. National Archives of Canada (hereafter NA), MG28-III, 1, Boyd Family Papers, vol. 875, file no. 1.
21. D. J. Wurtele, "Mossom Boyd: Lumber King of the Trent Valley," *Ontario History* 50 (Autumn 1958), 177–9. See also C. Grant Head, "An Introduction to Forest Exploitation in Nineteenth Century Ontario," in *Perspectives on Landscape and Settlement in Nineteenth Century Ontario*, ed. J. David Wood (Toronto: McClelland and Stewart, 1975), 78–112; Chris Curtis, "Shanty Life in the Kawarthas, 1850–1855," *Material History Bulletin/Bulletin d'histoire de la culture materielle* 13 (1981): 39–49; Lorne F. Hammond, "Anatomy of a Lumber Shanty: A Social History of Labour and Production on the Lievre River, 1876–1890," in *Canadian Papers in Rural History*, vol. 9, ed. Donald H. Akenson (Gananoque, ON: Langdale Press, 1994): 291–322.
22. Edwin C. Guillet, ed., *The Valley of the Trent* (Toronto: Champlain Society, 1957), 276.
23. Spragge, "Colonization Roads," 7–8.
24. AO, RG1, series A-I-7, Crown Land Department, Settlement Files [Bobcaygeon Road], vol. 12, env. 3, 6485, Richard Hughes Report, 15 January 1859; 6490, Hughes' Report, 31 December 1859 (both MS 892/reel 7).
25. AO, RG1, A-I-6, Crown Land Department Correspondence, vol. 31, env. 7, 27539–41, W. S. Conger to Richard Hughes, 20 February 1857 (MS 563/29).
26. Spragge, "Colonization Roads," 10.
27. AO, RG1, A-I-6, vol. 33, env. 4, file no. 13171, 28963–9, Robert McNaughton

to P. M. Vankoughnet, 8 October 1861 (MS 563/30). In this connection see Cecil J. Houston and William J. Smyth, *Irish Emigration and Canadian Settlement: Patterns, Links, and Letters* (Toronto: University of Toronto Press; Belfast: Ulster Historical Foundation, 1990), 62 – 3.

28. Ibid.
29. Ibid.
30. Ibid.
31. AO, RG1, A-I-6, vol. 33, env. 4, 28997 – 8, Richard Hughes to Vankoughnet, 24 October 1861 (MS 563/30). Cf. G.R.C. Keep, "A Canadian Emigration Commissioner in Northern Ireland," *Canadian Historical Review* 34 (June 1953): 151 – 7.
32. Jones, *History of Agriculture*, 291; Gates, *Land Policies*, 287.
33. AO, RG52, series 1-A, box 1, Colonization Roads, Bobcaygeon Road File, Hughes to William Hutton, 1 March 1858.
34. AO, RG52, 1-A, box 1, Mossom Boyd to Hutton, 5 March 1858.
35. AO, RG1, A-I-7, vol. 12, env. 3, 6485, 6490, Hughes' Report 31 December 1858 (MS 892/7).
36. AO, RG52, 1-A, box 1, Hughes to Vankoughnet, 31 December 1859; Hughes to Hutton, 1 October 1860.
37. AO, RG1, F-II-4, vol. 8, Crown Lands Department, Free Grant Timber Returns, 1861 – 1871.
38. Ibid.
39. Ibid.
40. NA, MG28-III, 1, vol. 137, Returns of Clearances, 1860 – 1861.
41. AO, RG1, A-I-7, vol. 12, env. 3, 6505 – 7, George Burwell to William McDougall [Commissioner of Crown Lands], 10 January 1863 (MS 892/7).
42. AO, RG1, A-I-7, vol. 12, env. 3, 6485, 6490, 6498, 6514, 6516, 6518, 6520, 6547, 6548, 6549, 6551, 6553, 6555, 6562 (MS 892/7); AO, RG52, 1-A, box 1, Pay List Nos. 1 and 2, 6 July; No. 3, 31 July; No. 4, 4 September; No. 5, 5 October; No. 6, 1 November; No. 7, 10 November; No. 1, 28 December 1866; No. 1, 12 August; No. 2, 7 October; Nos. 3 and 4, 28 October 1867.
43. Ibid.
44. Ibid.
45. Ibid.
46. Ibid.
47. NA, MG28-III, 1, vol. 138; vol. 106.
48. AO, RG1, A-I-7, vol. 12, env. 3, 6485, 6490, 6498, 6514, 6516, 6518, 6520, 6547, 6548, 6549, 6551, 6553, 6555, 6562 (MS 892/7); AO, RG52, 1-A, box 1, Pay List Nos. 1 and 2, 6 July; No. 3, 31 July; No. 4, 4 September; No. 5, 5 October; No. 6, 1 November; No. 7, 10 November; No. 1, 28 December 1866; No. 1, 12 August; No. 2, 7 October; Nos. 3 and 4, 28 October 1867.
49. Ontario, *Sessional Papers*, 1908, no. 3, "Report of the Minister of Lands, Forests and Mines of the Province of Ontario, 1907," 225, [citing Aubrey White, comp. "History of Crown Timber Regulations from the Date of the French Occupation to the Present Time," in *Annual Report of the Clerk of Forestry for Ontario*, 1899].
50. Ibid.
51. Ibid.
52. Ibid.
53. Ibid.
54. Canada, *Journals of the Legislative Assembly*, 1863, Appendix No. 8., Evidence of Alexander Dennistoun, 6 May 1863.
55. Canada, *Journals of the Legislative Assembly*, 1863, Appendix No. 8., Evidence of Allan Gilmour, 12 May 1863.
56. Canada, *Journals of the Legislative Assembly*, 1863, Appendix No. 8., Evidence of Robert A. Strickland, 7 May 1863.
57. Ibid.
58. Canada, *Journals of the Legislative Assembly*, 1863, Appendix No. 8., Evidence of Ezra Stephens, 7 May 1863.
59. Ibid.

60. Ibid.
61. Ibid. Stephens' assertion was corroborated by two men sceptical of the project and the accuracy of the provincial agents' and local boosters' reports. Respectively, see the letters by John Shearer and George Esson to the editor of the *Peterborough Examiner* 7 February 1861, 3 and 14 February 1861, 2 – 3.
62. Canada, *Journals of the Legislative Assembly*, 1863, Appendix No. 8., Evidence of Ezra Stephens, 7 May 1863.
63. Ibid.
64. AO, RG1, A-I-6, vol. 33, env. 5, 29100 – 1, J. Kirkland to G. H. Roche, 13 January 1862 (MS 563/31); vol. 34, env. 5, 30215 – 6, J. C. Singleton to Andrew Russell, 23 February 1865 (MS 563/32).
65. *Peterborough Examiner* 26 March 1863, 2.
66. NA, MG28-III, 1, vol. 94, Boyd to Norman Barnhart, 12 February 1870.
67. Canada, Commission of Conservation, *Trent Watershed Survey: A Reconnaissance*, compiled by C. D. Howe and J. H. White, with an introduction by B. E. Fernow (Toronto: Bryant Press, 1913), 95.
68. Charles Landy MacDermott, *Facts for Emigrants. A Journey from London to the Backwoods of Canada, Containing List of Places, Cost of Provisions, Information as to Distances, Wages, and Labour, Timber and Land, and Travelling Expenses from London to Haliburton, Township of Dysart, County of Peterborough, Canada West* (London: H. Born, 1868), 12 – 3.
69. Captain George S. Thompson, *Up to Date; Or, the Life of a Lumberman* (n.p.: Privately printed, 1895), 10.
70. Ibid.
71. Ibid.
72. Ibid., 10 – 1
73. Ibid.
74. J. I. Little, *Nationalism, Capitalism, and Colonization in Nineteenth-Century Quebec: The Upper St. Francis District* (Kingston and Montreal: McGill-Queen's University Press, 1989).
75. John Langton, "On the Age of Timber Trees and the Prospects of a Continuous Supply of Timber in Canada," *Transactions of the Literary and Historical Society of Quebec*, vol. 5., art. 3 (Quebec: G. T. Gary, 1862), 75 – 6.
76. Ibid.
77. Commission of Conservation, *Trent Watershed Survey*, 1 – 10. A province-wide history and assessment is rendered in B. E. Fernow, "Forest Resources and Forestry," in *Canada and Its Provinces: A History of the Canadian People and Their Institutions by One Hundred Associates*, vol. 18, sect. 9, part 2, *The Province of Ontario*, ed. Adam Shortt and Arthur G. Doughty (Toronto: Glasgow, Brook and Co., 1914), 585 – 99.
78. Commission of Conservation, *Trent Watershed Survey*, 5.
79. See H. V. Nelles, *The Politics of Development: Forests, Mines and Hydro-Electric Power in Ontario, 1849 – 1941*. (Toronto: University of Toronto Press, 1974); Robert Peter Gillis, "The Ottawa Lumber Barons and the Conservation Movement, 1880 – 1914," *Journal of Canadian Studies* 9 (February 1974): 14 – 30; Bruce W. Hodgins, Jamie Benidickson, and Peter Gillis, "The Ontario and Quebec Experiments in Forest Reserves, 1883 – 1930," *Journal of Forest History* 26 (January 1982): 20 – 33; Peter Gillis and Thomas R. Roach, *Lost Initiatives: Canada's Forest Industries, Forest Policy, and Forest Conservation* (Westport: Greenwood Press, 1986).
80. W. L. Morton, "Victorian Canada," in *The Shield of Achilles*, 330.

Chapter 6

1. Catharine Parr Traill, *Studies of Plant Life in Canada; Or, Gleanings from Forest, Lake, and Plain* (Ottawa: A. S. Woodburn, 1885), 2. Traill's story is told by G. H. Needler, *Otonabee Pioneers: The Story of the Stewarts, the Stricklands, the Traills, and the Moodies* (Toronto: Burns and MacEachern, 1953); Sara Eaton, *Lady of the Backwoods: A Biography of Catharine Parr Traill* (Toronto: McClelland and Stewart, 1969); and Charlotte Gray, *Sisters in the Wilderness: The Lives of Susanna Moodie and Catharine Parr Traill* (Toronto: Viking, 1999). Her work as an amateur scientist is explored by Michael Peterman "'Splendid Anachronism': The Record of Catharine Parr Traill's Struggles as an Amateur Botanist in Nineteenth-Century Canada," in *Re(Dis)covering Our Foremothers: Nineteenth-Century Canadian Women Writers*, ed. Lorraine McMullen (Ottawa: University of Ottawa Press, 1990), and Marianne Gosztonyi Ainley, "Science in Canada's Backwoods: Catharine Parr Traill," in *Natural Eloquence: Women Reinscribe Science*, ed. Barbara T. Gates and Ann B. Shteir (Madison: University of Wisconsin Press, 1997). For an introduction to nature writing in Canada see John Matthews, "Literature and Environment: Inheritance and Adaptation—The Canadian Experience," in *Commonwealth Literature: Unity and Diversity in a Common Culture*, ed. John Press (London: Heineman Education Books, 1964); Enid Mallory "Pioneer Naturalist, Catharine Parr Traill," *Canadian Audubon* 27 (March-April 1965): 42 – 5; Elaine Theberge, "The Untrodden Earth: Early Nature Writing in Canada," *Nature Canada* 3 (1974): 30 – 6; T. D. MacLulich, "Crusoe in the Backwoods: A Canadian Fable?" *Mosaic* 9 (Winter 1976): 115 – 26; T. D. MacLulich, "Reading the Land: The Wilderness Tradition in Canadian Letters," *Journal of Canadian Studies* 20 (Summer 1985): 29 – 44; Carole Gerson, "Nobler Savages: Representations of Native Women in the Writings of Susanna Moodie and Catharine Parr Traill," *Journal of Canadian Studies* 32 (Summer 1997): 5 – 21.
2. Peter Berg and Raymond Dasman, quoted in Doug Abberley, "Interpreting Bioregionalism: A Story from Many Voices," in *Bioregionalism*, ed. Michael Vincent McGinnis (London and New York: Routledge, 1999), 23.
3. Vera Norwood, *Made from this Earth: American Women and Nature* (Chapel Hill and London: University of North Carolina Press, 1993), 4. Cf. Ann B. Shteir, *Cultivating Women, Cultivating Science: Flora's Daughters and Botany in England 1760 to 1860* (Baltimore and London: The Johns Hopkins University Press, 1996), and Barbara T. Gates, *Kindred Nature: Victorian and Edwardian Women Embrace the Living World* (Chicago: University of Chicago Press, 1998).
4. A. B. McKillop, *A Disciplined Intelligence: Critical Inquiry and Canadian Thought in the Victorian Era* (Montreal: McGill-Queen's University Press, 1979); Carl Berger, *Science, God, and Nature in Victorian Canada* (Toronto: University of Toronto Press, 1983); Gerald Killan, *David Boyle: From Artisan to Archaeologist* (Toronto: University of Toronto Press with the Ontario Heritage Foundation, 1983); Suzanne Zeller, *Inventing Canada: Early Victorian Science and the Idea of a Transcontinental Nation* (Toronto: University of Toronto Press, 1987), esp. 195 – 6, 241; Suzanne Zeller, "The Spirit of Bacon: Science and Self-Perception in the Hudson's Bay Company," *Scientia*

Canadensis 13 (Fall/Winter 1989): 79–101; Susan Sheets-Pyenson, *Cathedrals of Science: The Development of Colonial Natural History Museums During the Late Nineteenth Century* (Kingston and Montreal: McGill-Queen's University Press, 1988); Susan Sheets-Pyenson, *John William Dawson: Faith, Hope, and Science* (Montreal and Kingston: McGill-Queen's University, 1996); W. A. Waiser, *The Field Naturalist: John Macoun, the Geological Survey, and Natural Science* (Toronto: University of Toronto Press, 1989); Paul A. Bogaard, ed., *Profiles of Science and Society in the Maritimes Prior to* 1914 (Sackville: Acadiensis Press and Centre for Canadian Studies, Mount Allison University, 1990); Nancy Christie, "Sir William Logan's Geological Empire and the 'Humbug' of Economic Utility," *Canadian Historical Review* 75 (June 1994): 161–204; cf. David Elliston Allen, *The Naturalist in Britain: A Social History* (London: Allen Lane, 1976). The period's social setting and intellectual contours are discussed by W. L. Morton, "Victorian Canada," in *The Shield of Achilles: Aspects of Canada in the Victorian Age*, ed. W. L. Morton (Toronto and Montreal: McClelland and Stewart, 1968), and Ramsay Cook, *The Regenerators: Social Criticism in Late Victorian English Canada* (Toronto: University of Toronto Press, 1985).

5. Donald Worster, *Nature's Economy: A History of Ecological Ideas* (Cambridge: Cambridge University Press, 1985), 1–55.
6. Ibid., 7–9.
7. Jonathan Bate, *Romantic Ecology: Wordsworth and the Environmental Tradition* (London: Routledge, 1991); Patricia Jasen, *Wild Things: Nature, Culture, and Tourism in Ontario, 1790–1914* (Toronto: University of Toronto Press, 1995), esp. 1–79; Franz K. Stanzel, "Innocent Eyes?: Canadian Landscape as Seen by Frances Brooke, Susanna Moodie, and Others," *International Journal of Canadian Studies/ Revue internationale d'etudes canadiennes* 4 (Fall/Automne 1991): 97–109; Barbara E. Wilson, "'Strangers in a Strange Land'—Literary Use of Canadian Landscape by Five Genteel Settlers," (master's thesis, University of Guelph, 1973). Wider contexts of the Romantic tradition and its bearing on national identity are explored in Larry Pratt and Matina Karvellas, "Nature and Nation: Herder, Myth and Cultural Nationalism in English Canada," *National History* 1 (Winter 1997): 59–77.
8. Catharine Parr Traill, "Floral Sketches, No. 1: The Violet [1843]," in *Forest and Other Gleanings: The Fugitive Writings of Catharine Parr Traill*, ed. Michael A. Peterman and Carl Ballstadt (Ottawa: University of Ottawa Press, 1995), 230.
9. Marian Fowler, *The Embroidered Tent: Five Gentlewomen in Early Canada* (Toronto: Anansi, 1982), 56–7, 62–3.
10. Catharine Parr Traill, *The Backwoods of Canada: Being Letters from the Wife of an Emigrant Officer, Illustrative of the Domestic Economy of British America* (London: Charles Knight, 1836; reprinted, Toronto: McClelland and Stewart, 1989), 55.
11. Ibid., 162.
12. Ibid., 163.
13. Ibid. See also David Lowenthal and Hugh C. Prince, "The English Landscape," *Geographical Review* 54 (July 1964): 309–46; "English Landscape Tastes," *Geographical Review* 55 (April 1965): 186–222.
14. Traill, *The Backwoods of Canada*, 102.
15. Ibid.
16. Ibid., 120.
17. Ibid., 209.
18. Ibid., 210.
19. Catharine Parr Traill, *Canadian Crusoes: A Tale of the Rice Lake Plains*, ed. Agnes

Strickland [ed. Rupert Schieder] (London: A. Hall, Virtue, 1852; reprinted, Ottawa: Carleton University Press, 1986).

20. Ibid., 209.
21. Ibid., 209 – 10.
22. Ibid., 210.
23. Ibid., 211.
24. Fuller treatments of British and Upper Canadian Indian affairs are R. J. Surtees, "The Development of an Indian Reserve Policy in Canada," *Ontario History* 61 (1969): 87 – 98; L.F.S. Upton, "The Origins of Canadian Indian Policy," *Journal of Canadian Studies* 8 (November 1973): 51 – 61; John Webster Grant, *Moon of Wintertime: Missionaries and the Indians of Canada in Encounter Since 1534* (Toronto: University of Toronto Press, 1984); Donald B. Smith, *Sacred Feathers: The Reverend Peter Jones (Kahkewaquonaby) and the Mississauga Indians* (Lincoln: University of Nebraska Press; and Toronto: University of Toronto Press, 1987).
25. Traill, *The Backwoods of Canada*, 119.
26. Catharine Parr Traill to William Traill, 1 October 1890, in *I Bless You in My Heart: Selected Correspondence of Catharine Parr Traill*, ed. Carl Ballstadt, Elizabeth Hopkins, and Michael A. Peterman (Toronto: University of Toronto Press, 1996), 341.
27. Ballstadt, Hopkins, and Peterman, ed., *I Bless You in My Heart*, 273 – 4.
28. Catharine Parr Traill, *Pearls and Pebbles; Or, Notes of an Old Naturalist* (Toronto: William Briggs, 1894), xxxii; Charles Pelham Mulvany, et al., *History of the County of Peterborough, Ontario; Containing a History of the County, History of Haliburton County, Their Townships, Towns, Schools, Churches; General and Local Statistics, Biographical Sketches, and an Outline History of the Dominion of Canada* (Toronto: C. Blackett Robinson, 1884), 218 – 22.
29. Traill to Sir Sanford Fleming, 21 December 1898, in Ballstadt, Hopkins, and Peterman, ed., *I Bless You in My Heart*, 392. Cf. Lee Clark Mitchell, *Witness to a Vanishing America: The Nineteenth-Century Response* (Princeton: Princeton University Press, 1981).
30. Catharine Parr Traill, "[The Rice Lake Plains, 1852] Forest Gleanings, No. 2," in Peterman and Ballstadt, ed., *Forest and Other Gleanings*, 200 – 1.
31. Catharine Parr Traill, "Ramblings by the River [1853]," in Peterman and Ballstadt, ed., *Forest and Other Gleanings*, 164.
32. Ibid., 166.
33. Ibid., 166.
34. Ibid.
35. Catharine Parr Traill, "A Walk to Railway Point [1853]," in Peterman and Ballstadt, ed., *Forest and Other Gleanings*, 216.
36. Traill, *Studies of Plant Life*, ii.
37. Ibid.
38. Ibid., 1.
39. Fowler, *The Embroidered Tent*, 65.
40. Traill, *Pearls and Pebbles*, 45 – 6.

Chapter 7

1. Major Samuel Strickland, *Twenty-Seven Years in Canada West; Or, the Experience of an Early Settler*, 2 vols., ed. Agnes Strickland (London: R. Bentley, 1853; reprinted, Edmonton: M. G. Hurtig Ltd., 1970), 2: 314.
2. Here I draw inspiration from Jay Gitlin, "On the Boundaries of Empire: Connecting the West to Its Imperial Past," in *Under an Open Sky: Rethinking America's Western Past*, ed. William Cronon, George Miles, and Jay Gitlin (New York: W. W. Norton, 1992), esp. 71. See also [Donald H. Akenson] "Pre-Confederation Canada as a European Society," in *Colonies: Canada*

to 1867, ed. David Jay Bercuson, et al. (Toronto: McGraw-Hill Ryerson, 1992), and Graeme Wynn, "On the Margins of Empire (1760 – 1840)," in *The Illustrated History of Canada*, ed. Craig Brown (Toronto: Key Porter Books, 1997).

3. Thomas R. Dunlap, *Nature and the English Diaspora: Environment and History in the United States, Canada, Australia, and New Zealand* (Cambridge and New York: Cambridge University Press, 1999), esp. 46 – 70. See also the collected essays in Tom Griffiths and Libby Robin, eds., *Ecology and Empire: Environmental History of Settler Societies* (Edinburgh: Keele University Press; Seattle: University of Washington Press, 1997).
4. Dan Flores, "Place: An Argument for Bioregional History," *Environmental History Review* 18 (Winter 1994): 1 – 18.
5. Thomas F. McIlwraith, *Looking for Old Ontario: Two Centuries of Landscape Change* (Toronto: University of Toronto Press, 1997); J. David Wood, *Making Ontario: Agricultural Colonization and Landscape Re-Creation Before the Railway* (Montreal and Kingston: McGill-Queen's University Press, 2000).
6. Peter Berg and Raymond Dasman, quoted in Doug Abberley, "Interpreting Bioregionalism: A Story from Many Voices," in *Bioregionalism*, ed. Michael Vincent McGinnis (London and New York: Routledge, 1999), 23.

◑

Bibliography

Abbreviations

Primary Documents

AO – Archives of Ontario, Toronto
NA – National Archives of Canada, Ottawa
TUA – Trent University Archives, Peterborough, Ontario

Secondary Journals

CHR – Canadian Historical Review
CPRH – Canadian Papers in Rural History
DCB – The Dictionary of Canadian Biography
EH – Environmental History
EHR– Environmental History Review
ER – Environmental Review
JCS – Journal of Canadian Studies
JFH – Journal of Forest History
OH – Ontario History

Manuscript Collections

Baird, (Nicol H.) Family Papers, AO.
Boyd, (Mossom) Family Papers, NA.
Huston, John Fonds, TUA.
Langton, John, AO.
Need, Thomas, AO.
Robinson, Peter Papers, AO.

Government Documents

AO. Record Group 1, Series A-I-6, Crown Lands Department, Correspondence, 1824–67.

AO. Record Group 1, Series A-I-7, vol. 10, Crown Lands Department, Subject Files, Mill Sites.

AO. Record Group 1, Series A-I-7, vols. 12, 16, Crown Lands Department, Subject Files, Settlement Files, 1826, 1859 – 66.

AO. Record Group 1, Series A-II-2, Crown Lands Department, Commissioners' Reports, Petition for Land Grants, 1832 – 35.

AO. Record Group 1, Series C-I-1, Crown Lands Department, Lands Branch, Applications, 1829, 1845.

AO. Record Group 1, Series C-IV, Crown Lands Department, Township Papers, 1829, 1858.

AO. Record Group 1, Series F-II-4, Vol. 8, Crown Lands Department, Free Grant Timber Returns, 1861 – 71.

AO. Record Group 22, Series 31, Cobourg, Newcastle District Court of General Quarter Sessions of the Peace, Case Files 1841 – 42.

AO. Record Group 22, Series 35, Cobourg, Newcastle District Court of General Quarter Sessions of the Peace, Road and Bridge Records, 1832.

AO. Record Group 22, Series 37, Cobourg, Newcastle District Court of General Quarter Sessions of the Peace, Petitions 1835 – 37.

AO. Record Group 52, Series 1-A, Box 1, Colonization Roads, Bobcaygeon Road File.

AO. Manuscript 180, Aggregate Census and Assessment Returns for Upper Canada, 1825–49.

Canada, Province of. *Journal of the Legislative Assembly*, 1841 – 67.

Canada, Province of. *Census of the Canadas*, 1851 – 2. Quebec: King's Printer, 1855.

Canada. *Census of the Canadas*, 1860 – 61. Quebec: S. B. Foote, 1863.

———. *Census of Canada*, 1870 – 71. Ottawa: I. B. Taylor, 1875.

———. *Indian Treaties and Surrenders. From 1680 to 1890*. 3 vols. Ottawa: Queen's Printer, 1891.

———. Commission of Conservation. *Trent Watershed Survey: A Reconnaissance*. Compiled by C. D. Howe and J. H. White. With an introduction by B. E. Fernow. Toronto: Bryant Press, 1913.

NA. Record Group 1, Series L 3, Upper Canada Land Petitions, 1775 – 95, 1806 – 08, 1833 – 35.

NA. Record Group 5, Series A 1, Civil Secretary's Correspondence, Upper Canada Sundries, 1816 – 40.

NA. Record Group 5, Series C 1, Canada West, Provincial Secretary's Correspondence, 1840 – 67.

NA. Record Group 43, Department of Railways and Canals, Canals Branch Correspondence, 1842.

Ontario. Department of Lands and Forest, Land and Survey Papers, 1825, 1849.

Ontario. "Report of the Minister of Lands, Forests and Mines of the Province of Ontario, 1907." *Sessional Papers*, 1908, no. 3.
Upper Canada. *Journal of the House of Assembly*, 1829 – 40.

Newspapers

British Whig (Kingston). 4 November 1834.
Christian Guardian (Toronto). 7 April 1847; 12 May 1847.
Chronicle and Gazette (Kingston). 27 December 1834.
Cobourg Star. 1831 – 49.
Colonial Advocate (York). 23 June 1831.
Peterborough Chronicle. 1845 – 46.
Peterborough Examiner. 1858 – 67.
Peterborough Gazette. 1845 – 48.
Peterborough Review. 1854 – 57
Reformer (Cobourg). 1832 – 37.
Toronto Constitution. 27 July 1836.
Upper Canada Herald (Kingston). 29 September 1830.
Weekly Despatch (Peterborough). 1846 – 56.

Primary Books

The Aborigines Protection Society. *Report on the Indians of Upper Canada*. London: William Ball, Arnold, and Company, 1839.
Drake, Daniel. *A Systematic Treatise, Historical, Etiological, and Practical, on the Principal Diseases of the Interior Valley of North America, as They Appear in the Caucasian, African, Indian, and Esquimaux Varieties of its Population*. In *Malaria in the Interior Valley of North America*, introduction by Norman D. Levine. Cincinnati: Winthrop B. Smith and Co.; Philadelphia: Grigg, Elliot and Co.; New York: Mason and Law, 1850; reprinted, Urbana: University of Illinois Press, 1964.
Hall, Basil. *Travels in North America in the Years 1827 and 1828*. 3 vols. Edinburgh: Cadell and Co.; London: Simpkin and Marshall, 1829.
MacDermott, Charles Landy. *Facts for Emigrants. A Journey from London to the Backwoods of Canada. Containing List of Places, Cost of Provisions, Information as to the Distances, Wages, and Labour, Timber and Land and Travelling Expenses from London to Haliburton, Township of Dysart, County Peterborough, Canada West*. London: H. Born, 1868.
Mulvany, Charles Pelham, et al. *History of the County of Peterborough, Ontario; Containing a History of the County, History of Haliburton County, Their Townships, Towns, Schools, Churches, etc., General and Local Statistics, Biographical Sketches, and an Outline History of the Dominion of Canada, etc.* Toronto: C. Blackett Robinson, 1884.

Need, Thomas. *Six Years in the Bush; Or, Extracts from the Journal of a Settler in Upper Canada.*.London: Simpkin, Marshall and Co., 1838.

Poole, Thomas W. *A Sketch of the Early Settlement and Subsequent Progress of the Town of Peterborough, and of Each Township in the County of Peterborough.* Peterborough, ON: Peterborough Review, 1867; reprinted, Peterborough, ON: Peterborough Printing Co., 1941.

Russell, Peter. *The Correspondence of the Honourable Peter Russell, with Allied Documents Relating to His Administration of the Government of Upper Canada During the Official Term of Lieut.-Governor J. G. Simcoe While on Leave of Absence.* 2 vols. Edited by E. A. Cruikshank and A. F. Hunter. Toronto: Ontario Historical Society, 1935.

Shirreff, Patrick. *A Tour through North America; Together with a Comprehensive View of the Canadians and the United States as Adopted for Agricultural Emigration.* Edinburgh: Oliver and Boyd, 1835.

Smith, William Loe. *The Pioneers of Old Ontario.* Toronto: George N. Morang, 1923.

Smith, W. H. *Canada: Past, Present and Future; Being a Historical, Geographical, Geological and Statistical Account of Canada West.* 2 vols. Toronto: Thomas Maclean, 1851.

Strickland, Major Samuel. *Twenty-Seven Years in Canada West: Or, the Experience of an Early Settler.* 2 vols. edited by Agnes Strickland. London: R. Bentley, 1853; reprinted, Edmonton: M. G. Hurtig, Ltd., 1970.

Thompson, Captain George S. *Up to Date; Or, the Life of a Lumberman.* n.p.: Privately printed, 1895.

Traill, Catharine Parr. *The Backwoods of Canada: Being Letters from the Wife of an Emigrant Officer, Illustrative of the Domestic Economy of British America.* London: Nattali and Bond, 1836.

———. *The Backwoods of Canada: Being Letters from the Wife of an Emigrant Officer, Illustrative of the Domestic Economy of British America.* London: Charles Knight, 1836; reprinted, Toronto: McClelland and Stewart, Inc., 1989.

———. *The Canadian Settler's Guide.* Toronto: Old Countryman Office, 1855; reprinted, Toronto: McClelland and Stewart, 1969.

———. *Canadian Crusoes: A Tale of the Rice Lake Plains.* Edited by Agnes Strickland. London: A. Hall, Virtue, 1852; reprinted, Ottawa: Carleton University Press, 1986.

———. *Studies of Plant Life in Canada; Or, Gleanings from Forest, Lake, and Plain.* Ottawa: A. S. Woodburn, 1885.

———. *Pearls and Pebbles; Or, Notes of an Old Naturalist.* Toronto: William Briggs, 1894.

Weld, Charles Richard. *A Vacation Tour in the United States and Canada.* London: Longman, Brown, Green, and Longmans, 1854.

White, Thomas. *An Exhibit of the Progress, Position and Resources, of the County of Peterboro, Canada West, Based Upon the Census of 1861; Together with a Statement of the Trade of the Town of Peterborough.* Peterborough, ON: T & R White, 1862.

Primary Articles or Chapters

Chamberlain, A. F. "Notes on the History, Customs, and Beliefs of the Mississaugas." *Journal of American Folklore* 1 (July 1888): 150 – 60.

Fernow, B. E. "Forest Resources and Forestry." In *The Province of Ontario*, edited by Adam Shortt and Arthur G. Doughty, vol. 18, sect. 9, part 2, *Canada and Its Provinces: A History of the Canadian People and Their Institutions by One Hundred Associates*. Toronto: Glasgow, Brook and Company, 1914.

Langton, John. "On the Age of Timber Trees and the Prospects of a Continuous Supply of Timber in Canada." *Transactions of the Literary and Historical Society of Quebec* vol. 5, art. 3, 61 – 79. Quebec: G. T. Gary, 1862.

[Traill, Catharine Parr]. "Flowers and Their Moral Teaching." *British American Magazine* 1 (May 1863): 55 – 9.

Secondary Books

Allen, David Elliston. *The Naturalist in Britain: A Social History*. London: Allen Lane, 1976.

Anderson, James E. *The Serpent Mounds Site Physical Anthropology*. Occasional Paper 11. Art and Archaeology, Royal Ontario Museum. University of Toronto. Toronto: University of Toronto Press, 1968.

Angus, James T. *A Respectable Ditch: A History of the Trent Canal, 1833 – 1920*. Kingston and Montreal: McGill-Queen's University Press, 1988.

Ballstadt, Carl, Elizabeth Hopkins, and Michael A. Peterman, eds. *I Bless You in My Heart: Selected Correspondence of Catharine Parr Traill*. Toronto: University of Toronto Press, 1996.

Bate, Jonathan. *Romantic Ecology: Wordsworth and the Environmental Tradition*. London: Routledge, 1991.

Bell, Jonathan, and Mervyn Watson. *Irish Farming: Implements and Techniques*. Edinburgh: John Donald, Ltd., 1986.

Berger, Carl. *The Sense of Power: Studies in the Ideas of Canadian Imperialism, 1867 –1914*. Toronto: University of Toronto Press, 1970.

———. *The Writing of Canadian History: Aspects of English-Canadian Historical Writing Since 1900*. Toronto: Oxford University Press, 1976; reprinted, Toronto: University of Toronto Press, 1986.

———. *Science, God, and Nature in Victorian Canada*. Toronto: University of Toronto Press, 1983.

Bogaard, Paul, ed. *Profiles of Science and Society in the Maritimes Prior to 1914*. Sackville, NB: Acadiensis Press and the Centre for Canadian Studies, Mount Allison University, 1990.

Careless, J.M.S. *Canada: A Story of Challenge*. Rev. ed. Toronto: Macmillan of Canada, 1970.

Chapman, L. J., and D. F. Putnam. *The Physiography of Southern Ontario*. 2d ed. Toronto: University of Toronto Press, for the Ontario Research Foundation, 1966.

Clark, Andrew H. *The Invasion of New Zealand by People, Plants, and Animals: The South Island*. New Brunswick, NJ: Rutgers University Press, 1949.

———. *Three Centuries and the Island: A Historical Geography of Settlement and Agriculture in Prince Edward Island*. Toronto: University of Toronto Press, 1959.

———. *Acadia: The Geography of Nova Scotia to 1760*. Madison: University of Wisconsin Press, 1968.

Cole, Jean Murray. *David Fife and Red Fife Wheat: The Story of a Few Grains of Wheat that Were Saved From a Cow*. Keene, ON: Lang Pioneer Village, 1992.

Cook, Ramsay. *The Regenerators: Social Criticism in Lat Victorian English Canada*. Toront: University of Toronto Press, 1985.

Cowan, Helen. *British Emigration to British North America, 1783 – 1837*. Toronto: University of Toronto, 1928.

Craig, Gerald M. *Upper Canada: The Formative Years, 1784 – 1841*. Toronto: McClelland and Stewart, 1963.

Creighton, Donald. *The Commercial Empire of the St. Lawrence*. Toronto: University of Toronto Press, 1937.

Cronon, William. *Changes in the Land: Indians, Colonists, and the Ecology of New England*. New York: Hill and Wang, 1983.

———. *Nature's Metropolis: Chicago and the Great West*. New York: Norton, 1991.

Crosby, Alfred W. *The Columbian Exchange: Biological and Cultural Consequences of 1492*. Westport, CT: Greenwood Press, 1972.

———. *Ecological Imperialism: The Biological Expansion of Europe, 900 – 1900*. Cambridge and New York: Cambridge University Press, 1986.

Cummings, H. R. *Early Days in Haliburton*. Toronto: Ryerson University Press, for the Ontario Department of Lands and Forests, 1962.

Donnelly, James S., Jr. *The Land and the People of Nineteenth-Century Cork: The Rural Economy and the Land Question*. London: Routledge and Kegan Paul, 1975.

Dunlap, Thomas R. *Nature and the English Diaspora: Environment and History in the United States, Canada, Australia, and New Zealand*. Cambridge and New York: Cambridge University Press, 1999.

Eaton, Sara. *Lady of the Backwoods: A Biography of Catharine Parr Traill*. Toronto: McClelland and Stewart, 1969.

Farmer, Samuel. *On the Shores of Scugog*. Port Perry, ON: Port Perry Star, 1934.

Fowler, Marian. *The Embroidered Tent: Five Gentlewomen in Early Canada*. Toronto: Anansi, 1982.

Freeman, T. W. *Ireland: Its Physical, Historical, Social and Economic Geography*. London: Methuen and Co.; New York: E. P. Dutton and Co., 1950.

Gaffield, Chad, and Pam Gaffield, eds. *Consuming Canada: Readings in Environmental History*. Toronto: Copp-Clark, 1995.

Gates, Barbara T. *Kindred Nature: Victorian and Edwardian Women Embrace the Living World*. Chicago: University of Chicago Press, 1998.

Gates, Lillian F. *Land Policies of Upper Canada*. Toronto: University of Toronto Press, 1968.

Gilles, Herbert M., and David A. Warrell, eds. *Bruce-Chwatt's Essential Malariology*. 3d ed. London: Edward Arnold, 1993.

Gillis, R. P., and Thomas Roach. *Lost Initiatives: Canada's Forest Industries, Forest Policy, and Forest Conservation*. New York: Greenwood Press, 1986.

Grant, John Webster. *Moon of Wintertime: Missionaries and the Indians of Canada in Encounter Since 1534*. Toronto: University of Toronto Press, 1984.

Gray, Charlotte. *Sisters in the Wilderness: The Lives of Susanna Moodie and Catharine Parr Traill*. Toronto: Viking, 1999.

Greer, Alan. *The Patriots and the People: The Rebellion of 1837 in Rural Lower Canada*. Toronto: University of Toronto Press, 1993.

Griffith, Tom, and Libby Robin, eds. *Ecology and Empire: Environmental History of Settler Societies*. Edinburgh: Keele University Press; Seattle: University of Washington Press, 1997.

Grove, Richard H. *Green Imperialism: Colonial Expansion, Tropical Edens and the Origins of Environmentalism, 1600 – 1860*. Cambridge and New York: Cambridge University Press, 1995.

Guillet, Edwin C., ed. *The Valley of the Trent*. Toronto: Champlain Society, 1957.

Helmuth, Hermann. *The Quakenbush Skeleton: Osteology and Culture*. Trent University Occasional Papers in Anthropology, no. 9. Peterborough: Trent University, 1993.

Horwitz, Morton. *The Transformation of American Law, 1780 – 1860*. Cambridge, MA: Harvard University Press, 1977.

Houston, Cecil. *The Orange Order in Nineteenth Century Ontario: A Study in Institutional Cultural Transfer*. Toronto: University of Toronto Press, 1974.

Houston, Cecil, and William J. Smyth. *The Sash Canada Wore: A Historical Geography of the Orange Order in Canada*. Toronto: University of Toronto Press, 1980.

———. *Irish Emigration and Canadian Settlement: Patterns, Links, and Letters*. Toronto: University of Toronto Press; Belfast: Ulster Historical Foundation, 1990.

Hunter, Louis C. *A History of Industrial Power in the United States*. Vol. 1, *Waterpower in the Century of the Steam Engine*. Charlottesville: University Press of West Virginia, 1979.

Innis, Harold A. *The Fur Trade in Canada: An Introduction to Canadian Economic History*. New Haven: Yale University Press, 1930.

———. *The Cod Fisheries: The History of an International Economy*. New Haven: Yale University Press, 1940.

Jasen, Patricia. *Wild Things: Nature, Culture and Tourism in Ontario, 1790 – 1914*. Toronto: University of Toronto Press, 1995.

Johnston, H.J.M. *British Emigration Policy, 1815 – 1830: Shovelling Out Paupers*. Oxford: Clarendon Press, 1972.

Johnston, Richard B. *The Archaeology of the Serpent Mounds Site*. Occasional Paper 10. Art and Archaeology, Royal Ontario Museum. University of Toronto. Toronto: University of Toronto Press, 1968.

———. *Archaeology of Rice Lake, Ontario*. Anthropology Papers, no. 19. Ottawa: National Museum of Canada, 1968.

Jones, Robert L. *History of Agriculture in Ontario, 1613 – 1880*. Toronto: University of Toronto Press, 1946; reprinted, Toronto: University of Toronto Press, 1977.

Judd, Richard W. *Aroostook: A Century of Logging in Northern Maine*. Orono: University of Maine Press, 1989.

Keeney, Elizabeth B. *The Botanizers: Amateur Scientists in Nineteenth-Century America*. Chapel Hill and London: University of North Carolina Press, 1992.

Killan, Gerald. *David Boyle: From Artisan to Archaeologist*. Toronto: University of Toronto Press with the Ontario Heritage Foundation, 1983.

Kirkconnell, Watson. *County of Victoria Centennial History*. Rev. ed. With research assistance from Frankie L. MacArthur. Lindsay, ON: John Deyell Ltd., 1967.

Knorr, Klaus E. *British Colonial Theories, 1570 – 1850*. Toronto: University of Toronto Press, 1944.

Langton, W. A. *Early Days in Upper Canada, Letters of John Langton from the Backwoods of Upper Canada and the Audit Office of the Province of Canada*. Toronto: Macmillan, 1926.

Lambert, Richard S., with Paul Pross. *Renewing Nature's Wealth: A Centennial History of the Public Management of Lands, Forests, and Wildlife in Ontario, 1763 –1967*. Toronto: Hunter and Rose, Co. for the Ontario Department of Lands and Forests, 1967.

Legget, Robert. *Rideau Waterway*. Toronto University of Toronto Press, 1955.

Leung, Felicity L. *Grist and Flour Mills in Ontario: From Millstones to Rollers, 1780s–1880s*. History and Archaeology Series no. 53. Ottawa: Parks Canada, 1981.

Little, J. I. *Nationalism, Capitalism and Colonization in Nineteenth-Century Quebec: The Upper St. Francis District*. Kingston and Montreal: McGill-Queen's University Press, 1989.

Lower, Arthur R. M. *The North American Assault on the Canadian Forest: A History of the Lumber Trade Between Canada and the United States*. Toronto: Ryerson Press, 1938.

———. *Great Britain's Woodyard: British America and the Timber Trade, 1763 – 1867*. Montreal: McGill-Queen's University Press, 1973.

MacLennan, Hugh. *Seven Rivers of Canada*. Toronto: Macmillan of Canada, 1961.

Mannion, John. *Irish Settlements in Eastern Canada: A Study of Cultural Transfer and Adaptation*. Toronto: University of Toronto Press, 1974.

McCalla, Douglas. *Planting the Province: The Economic History of Upper Canada, 1784 – 1870*. Toronto: University of Toronto Press, 1993.

McGinnis, Michael Vincent, ed. *Bioregionalism*. London and New York: Routledge, 1999.

McIlwraith, Thomas F. *Looking for Old Ontario: Two Centuries of Landscape Change*. Toronto: University of Toronto Press, 1997.

McKillop, A. B. *A Disciplined Intelligence: Critical Inquiry and Canadian Thought in the Victoria Era* Montreal: McGill-Queen's University Press, 1979.

McLean, Marianne. *The People of Glengarry: Highlanders in Transition, 1745 – 1820*. Kingston and Montreal: McGill-Queen's University Press, 1991.

Melville, Elinor G. K. *A Plague of Sheep: Environmental Consequences of the Conquest of Mexico*. Cambridge and New York: Cambridge University Press, 1994.

Merchant, Carolyn. *Ecological Revolutions: Nature, Gender, and Science in New England*. Chapel Hill and London: University of North Carolina Press, 1989.

Morgan, Edmund S. *Inventing the People: The Rise of Popular Sovereignty in England and America*. New York: W. W. Norton, 1988.

Morton, W. L. *Manitoba: A History*. Toronto: University of Toronto Press, 1957.

Needler, G. H. *Otonabee Pioneers: The Story of the Stewarts, the Stricklands, the Traills, and the Moodies*. Toronto: Burns and MacEachern, 1953.

Nelles, H. V. *The Politics of Development: Forests, Mines and Hydro-Electric Power in Ontario, 1849 – 1941*. Toronto: University of Toronto Press, 1974.
O'Flanagan, Patrick, and Cornelius G. Buttimer, eds. *Cork History and Society: Interdisciplinary Essays on the History of an Irish County*. Dublin: Geography Publications, 1993.
Owram, Doug. *Promise of Eden: The Canadian Expansionist Movement and the Idea of the West, 1856 – 1900*. Toronto: University of Toronto Press, 1980; reprinted, Toronto: University of Toronto Press, 1992.
Palmer, Bryan D. *Working-Class Experience: Rethinking the History of Canadian Labour, 1800 – 1991*. 2d ed. Toronto: McClelland and Stewart, 1992.
Pammett, Howard T. *Lilies and Shamrocks: A History of Emily Township, County of Victoria, Ontario, 1818- 1973*. Lindsay, ON: John Deyell Co., 1974.
Peterman, Michael A., and Carl Ballstadt, eds. *Forest and Other Gleanings: The Fugitive Writings of Catharine Parr Traill*. Ottawa: University of Ottawa Press, 1994.
Pyne, Stephen J. *Fire in America: A Cultural History of Wildland and Rural Fire*. Princeton: Princeton University Press, 1982.
———. *Burning Bush: A Fire History of Australia*. New York: Henry Holt and Co., 1991.
Reaman, George Elmore. *A History of Vaughan Township: Two Centuries of Life in the Township*. Toronto: C. H. Snider, 1971.
Richie, William A. *An Archaeological Survey of the Trent Waterway in Ontario, Canada and its Significance for New York State Prehistory*. Research Records of the Rochester Museum of Arts and Sciences, no. 9. Rochester: Rochester Museum of Arts and Sciences, 1949.
See, Scott. *Riots in New Brunswick: Orange Nativism and Social Violence in the 1840s*. Toronto: University of Toronto Press, 1993.
Senior, Hereward. *Orangeism: The Canadian Phase*. Toronto: McGraw-Hill Ryerson, 1972.
Susan Sheets-Pyenson, *Cathedrals of Science: The Development of Colonial Natural History Museums During the Late Nineteenth Century*. Kingston and Montreal: McGill-Queen's University Press, 1988
———. *John William Dawson: Faith, Hope, and Science*. Montreal and Kingston: McGill-Queen's University, 1996.
Shteir, Ann B. *Cultivating Women, Cultivating Science: Flora's Daughters and Botany in England 1760 to 1860*. Baltimore and London: The Johns Hopkins University Press, 1996.
Skelton, Isabel. *The Backwoodswoman: A Chronicle of Pioneer Home Life in Upper and Lower Canada*. Toronto: Ryerson Press, 1923.
Smith, Donald B. *Sacred Feathers: The Reverend Peter Jones (Kahkewaquonaby) and the Mississauga Indians*. Lincoln: University of Nebraska Press; Toronto: University of Toronto Press, 1987.
Smith, William Loe. *The Pioneers of Old Ontario*. Toronto: G. N. Morang, 1923.
Spence, Michael W., and J. Russell Harper. *The Cameron's Point Site*. Occasional Paper 12. Art and Archaeology, Royal Ontario Museum. University of Toronto. Toronto: University of Toronto Press, 1968.
Steinberg, Theodore, *Nature Incorporated: Industrialization and the Waters of New England*. Cambridge and New York: Cambridge University Press, 1991.

Tahiliani, Moti, ed. *"Bless These Walls": Lindsay's Heritage*. Lindsay, ON: Blewet Printing, 1982.
Taylor, Alan. *William Cooper's Town: Power and Persuasion on the Frontier of the Early American Republic*. New York: Vintage, 1995.
Theberge, Clifford B., and Elaine Theberge. *At the Edge of the Shield: A History of Smith Township, 1818–1980*. Peterborough, ON: Alger Press, Ltd., for the Smith Township Historical Committee, 1982.
Vastokas, Joan M., and Romas K. Vastokas. *Sacred Art of the Algonkians: A Study of the Peterborough Petroglyphs*. Peterborough, ON: Manasard Press, 1973.
Weir, F. G. *Scugog and its Environs*. Port Perry, ON: Star Prints, 1927.
White, Philip. *Beekmantown, New York: Forest Frontier to Farm Community*. Austin: University of Texas Press, 1979.
Williams, Michael. *Americans and Their Forests: A Historical Geography*. Cambridge: Cambridge University Press, 1989.
Williams, Raymond. *Problems in Materialism and Culture: Selected Essays*. London: Verso, 1980.
Wilton, Carol. *Popular Politics and Political Culture in Upper Canada, 1800–1850*. Montreal: McGill-Queen's University Press, 2000.
Wood, J. David. *Making Ontario: Agricultural Colonization and Landscape Re-Creation Before the Railway*. Montreal and Kingston: McGill-Queen's University Press, 2000.
Worster, Donald. *Nature's Economy: A History of Ecological Ideas*. San Francisco: Sierra Club Books, 1977; reprinted, Cambridge and New York: Cambridge University Press, 1985.
Wynn, Graeme. *Timber Colony: A Historical Geography of Early Nineteenth Century New Brunswick*. Toronto: University of Toronto Press, 1981.
Zeller, Suzanne. *Inventing Canada: Early Victorian Science and the Idea of a Transcontinental Nation*. Toronto: University of Toronto Press, 1987.

Secondary Articles and Chapters

Aberley, Doug. "Interpreting Bioregionalism: A Story from Many Voices." In *Bioregionalism*, edited by Michael Vincent McGinnis. London and New York: Routledge, 1999.
Adams, David P. "Malaria, Labor, and Population Distribution in Costa Rica: A Biohistorical Perspective." *Journal of Interdisciplinary History* 27 (Summer 1996): 75–86.
Adams, W. P. "Notes on the Climates of Peterborough." In *The Geography of the Peterborough Area*. Peterborough, ON: Trent University, Department of Geography, 1972.
———. "The Climate of Peterborough and the Kawarthas: Past and Present." In *Peterborough and the Kawarthas*, edited by Peter Adams and Colin Taylor. Peterborough, ON: Heritage Publications, 1985.

Ainley, Marianne Gosztonyi. "Science in Canada's Backwoods: Catharine Parr Traill." In *Natural Eloquence: Women Reinscribe Science*, edited by Barbara T. Gates and Ann B. Shteir. Madison: University of Wisconsin Press, 1997.

Akenson, Donald H. "Pre-Confederation Canada as a European Society." In *Colonies: Canada to 1867*, edited by David Jay Bercuson, et al. Toronto: McGraw-Hill, 1992.

Armstrong, F. H. "The York Riots of March 23, 1832." *OH* 55 (June 1963): 61 – 72.

Birdwell-Pheasant, Donna. "The Home 'Place': Center and Periphery in Irish House and Family Systems." In *House Life: Space, Place and Family in Europe*, edited by Donna Birdwell-Pheasant and Denise Lawrence-Zuniga. Oxford and New York: Berg, 1999.

Blair, Blair. "Taken for 'Granted': Aboriginal Title and Public Fishing Rights in Upper Canada." *Ontario History* 92 (Spring 2000): 31 – 55.

Bleasdale, Ruth. "Class Conflict on the Canals of Upper Canada in the 1840s." *Labour/Le Travailleur* 7 (1981): 9 – 89.

Boyd, Robert T. "Another Look at the 'Fever and Ague' of Western Oregon." *Ethnohistory* 22 (Spring 1975): 135 – 54.

Brown, Quentin. "Swinging with the Governors: Newcastle District Election, 1836 and 1841." *OH* 86, 4 (December 1994): 319 – 36.

Brunger, Alan G. "Geographical Propinquity Among Pre-Famine Catholic Irish Settlers in Upper Canada." *Journal of Historical Geography* 8, 3 (1982): 265 – 82.

Buttle, Jim. "The Kawartha Lakes." In *Peterborough and the Kawarthas*, edited by Peter Adams and Colin Taylor. Peterborough, ON: Heritage Publications, 1985.

Cadigan, Sean. "Paternalism and Politics: Sir Francis Bond Head, the Orange Order, and the Election of 1836." *CHR* 72 (September 1991): 319 – 47.

Cameron, Wendy. "McDonell, Alexander." In *Dictionary of Canadian Biography*, vol. 9, 1861 – 70. Toronto: University of Toronto Press, 1976.

———. "Selecting Peter Robinson's Irish Emigrants." *Historie sociale-Social History* 9 (May 1976): 29 – 46.

———. "Nicol Hugh Baird and the Construction of the Trent-Severn Waterway." In *CPRH*. Vol. 7, edited by Donald H. Akenson. Gananoque, ON: Langdale Press, 1990.

Careless, J.M.S. "Fronterism, Metropolitanism, and Canadian History." *CHR* 35 (March 1954): 1 – 21.

———. "'Limited Identities' in Canada." *CHR* 50 (March 1969): 1 – 10.

Catling, P. M., V. R. Catling, and S. M. McKay-Kuja. "The Extent, Floristic Composition and Maintenance of the Rice Lake Plains, Ontario, Based on Historical Records." *Canadian Field-Naturalist* 106 (January-March 1992): 73 – 86.

Catling, P. M., and V. R. Catling. "Floristic Composition, Phytogeography and Relationships of Prairies, Savannas and Sand Barrens Along the Trent River, Eastern Ontario." *Canadian Field-Naturalist* 107 (January-March 1993): 24 – 45.

Christie, Nancy. "Sir William Logan's Geological Empire and the 'Humbug' of Economic Utility." *CHR* 75 (June 1994): 161 – 204.

Clarke, Aidan, with R. Dudley Edwards. "Pacification, Plantation, and the Catholic Question, 1603 – 23." In *A New History of Ireland*. Vol. 3, *Early Modern Ireland, 1534 – 1691*, edited by T. W. Moody, F. X. Martin, and F. J. Byrne. Oxford: Clarendon Press, 1976.

Connell, K. H. "The Colonization of Waste Land in Ireland, 1780 – 1845." *Economic History Review* 2d ser., 3, 1 (1950): 44 – 71.
Cook, Ramsay. "Canadian Centennial Celebrations." *International Journal* 22 (Autumn 1967): 659 – 63.
———. "Cabbages Not Kings: Towards an Ecological Interpretation of Early Canadian History." *JCS* 24 (Winter 1990 – 91): 5 – 16.
Craig, Beatrice. "Agriculture and the Lumberman's Frontier in the Upper St. John Valley, 1800 – 70." *JFH* 32 (July 1988): 125 – 37.
Cross, Michael S. "Stony Monday, 1849: The Rebellion Losses Riots in Bytown." *OH* 63 (September 1971): 175 – 90.
———. "The Shiners's War: Social Violence in the Ottawa Valley in the 1830s." *CHR* 54 (March 1973): 1 – 26.
———. "'The Laws are Like Cobwebs': Popular Resistance to Authority in Mid-Nineteenth Century North America." In *Law in a Colonial Society: The Nova Scotia Experience*, edited by Peter B. Waite, Sandra Oxner, and Thomas Barnes. Toronto: Carswell, 1984.
Crowley, Terence. "'Thunder Gusts': Popular Disturbances in Early French Canada." Canadian Historical Association, *Historical Papers* (1979): 11 – 32.
Crozier, M. J. "Geology and Geomorphology." In *The Geography of the Peterborough Area*. Peterborough, ON: Trent University, Department of Geography, 1972.
Curtis, Chris. "Shanty Life in the Kawarthas, 1850 – 1855." *Material History Bulletin/Bulletin d'histoire de la culture materielle* 13 (1981): 39 – 49.
Dickson, David. "Butter Comes to Market: The Origins of Commercial Dairying in County Cork." In *Cork History and Society: Interdisciplinary Essays on the History of an Irish County*. Dublin: Geography Publications, 1993.
Donahue, Brian, "'Damned at Both Ends and Cursed in the Middle': The 'Flowage' of the Concord River Meadows, 1798 – 1862." *ER* 13 (Fall/Winter 1989): 47 – 67.
Dunn, Fredrick L. "Malaria." In *The Cambridge World History of Human Disease*, edited by Kenneth F. Kiple. Cambridge: Cambridge University Press, 1993.
Ecclestone, Miles. "The Physical Landscape of Peterborough and the Kawarthas." In *Peterborough and the Kawarthas*, edited by Peter Adams and Colin Taylor. Peterborough, ON: Heritage Publications, 1985.
Ennals, Peter. "Cobourg and Port Hope: The Struggle for the 'Back Country.'" In *Perspectives on Landscape and Settlement in Nineteenth Century Ontario*, edited by J. David Wood. Ottawa: Carleton University Press, 1975.
Fallis, A. Murray. "Malaria in the 18th and 19th Centuries in Ontario." *Bulletin canadien d'historie de la medicine/Canadian Bulletin of Medical History* 1 (Hiver/Winter 1984): 25 – 38.
Fallis, Laurence, Jr. "The Idea of Progress in the Province of Canada: A Study in the History of Ideas." In *The Shield of Achilles: Aspects of Canada in the Victorian Age*, edited by W. L. Morton. Toronto: McClelland and Stewart, 1968.
Fitzpatrick, David. "Emigration, 1801 – 70." In *A New History of Ireland*. Vol. V, edited by T. W. Moody, F. X. Martin, and F. J. Byrne. Oxford: Clarendon Press, 1976
Flores, Dan. "Place: An Argument for Bioregional History." *EHR* 18 (Winter 1994): 1 – 18.

Forkey, Neil S. "Damning the Dam: Ecology and Community in Ops Township, Upper Canada." *CHR* 79 (March 1998): 68 – 99.

Gates, Paul W. "Official Encouragement to Immigration by the Province of Canada." *CHR* 15 (March 1934): 24 – 38.

George, Peter, and Philip Sworden. "The Courts and the Development of Trade in Upper Canada, 1830 – 1860." *Business History Review* 60 (Summer 1986): 258 – 80.

———. "John Beverly Robinson and the Commercial Empire of the St. Lawrence." *Research in Economic History* 11 (1988): 217 – 42.

Gerson, Carole. "Nobler Savages: Representations of Native Women in the Writings of Susanna Moodie and Catharine Parr Traill." *JCS* 32 (Summer 1997): 5 – 21.

Gitlin, Jay. "On the Boundaries of Empire: Connecting the West to Its Imperial Past." In *Under an Open Sky: Rethinking America's Western Past*, edited by William Cronon, George Miles, and Jay Gitlin. New York: W. W. Norton, 1992.

Hammond, Lorne F. "Anatomy of a Lumber Shanty: A Social History of Labour and Production on the Lievre River, 1876 – 1890." In *CPRH*. Vol. 9, edited by Donald H. Akenson. Gananoque, ON: Langdale Press, 1994.

Harris, R. Cole, Pauline Roulson, and Chris De Freitas. "The Settlement of Mono Township." *Canadian Geographer* 19, 1 (1975): 1 – 17.

Harris, Cole. "Regionalism and the Canadian Archipelago." In *Heartland and Hinterland: A Regional Geography of Canada*, 3d ed., edited by Larry McCann and Angus Gunn. Scarborough: Prentice Hall Canada, 1998.

Head, C. Grant. "An Introduction to Forest Exploitation in Nineteenth-Century Ontario." In *Perspectives on Landscape and Settlement in Nineteenth Century Ontario*, edited by J. David Wood. Ottawa: Carleton University Press, 1975.

Helleiner, Fred. "The Biogeography of the Peterborough Region." In *Peterborough and the Kawarthas*, edited by Peter Adams and Colin Taylor. Peterborough, ON: Heritage Publications, 1985.

Hodgins, Bruce W., Jamie Benidickson, and Peter Gillis. "The Ontario and Quebec Experiments in Forest Reserves, 1883 – 1930." *JFH* 26 (January 1982): 20 – 33.

Johnson, J. K. "'Claims of Equity and Justice': Petitions and Petitioners in Upper Canada, 1815 – 1840." *Historie sociale/Social History* 28 (May 1995): 219 – 40.

Johnson, Leo. "The Gore District 'Outrages', 1826 – 1829: A Case Study in Violence, Justice, and Political Propoganda." *OH* 83 (June 1991): 107 – 26.

Jones, Elwood. "Purdy, William." In *DCB*, vol. 7, 1836 – 50. Toronto: University of Toronto Press, 1988.

Kealey, Gregory S. "Orangemen and the Corporation." In *Forging a Consensus: Historical Essays on Toronto*, edited by Victor L. Russell. Toronto: University of Toronto Press, 1984.

Keep, G.R.C. "A Canadian Emigration Commissioner in Northern Ireland." *CHR* 34 (June 1953): 151 – 7.

Kenny, Stephen. "'Cahots' and Catcalls: An Episode of Popular Resistance in Lower Canada at the Outset of the Union." *CHR* 65 (June 1984): 184 – 208.

Kulik, Gary. "Dams, Fish, and Farmers: Defense of Public Rights in Eighteenth-Century Rhode Island." In *The Countryside in the Age of Capitalist Transformation*, edited by Steven Hahn and Jonathan Prude. Chapel Hill: University of North Carolina Press, 1985.

Lewthwaite, Susan. "Violence, Law, and Community in Rural Upper Canada." In *Essays in the History of Canadian Law*. Vol. 5., *Crime and Criminal Justice in Canadian History*, edited by Jim Phillips, Tima Loo, and Susan Lewthwaite. Toronto: University of Toronto Press and the Osgoode Society for Canadian Legal History, 1994.

Lowenthal, David, and Hugh C. Prince. "The English Landscape." *Geographical Review* 54 (July 1964): 309 – 46.

———. "English Landscape Tastes." *Geographical Review* 55 (April 1965): 186 – 222.

Lower, A.R.M. "Settlement and the Forest Frontier in Eastern Canada." In *Canadian Frontiers of Settlement*. Vol. 9, edited by W. A. Mackintosh and W.L.G. Joerg. Toronto: University of Toronto Press, 1936.

MacLulich, T. D. "Crusoe in the Backwoods: A Canadian Fable?" *Mosaic* 9 (Winter 1976): 115 – 26.

———. "Reading the Land: The Wilderness Tradition in Canadian Letters." *JCS* 20 (Summer 1985): 29 – 44.

Macoun, W. T. "Why Our Field and Roadside Weeds Are Introduced Species." *Ottawa Naturalist* 19, 6 (1905): 124 – 5.

MacNamara, Charles. "Champlain as a Naturalist." *Canadian Field-Naturalist* 40, 6 (1926): 125 – 33.

Mallory, Enid. "Pioneer Naturalist, Catharine Parr Traill." *Canadian Audubon* 27 (March-April 1965): 42 – 5.

Maltby, Peter L., and Monica Maltby, "A New Look at the Peter Robinson Emigration of 1823." *Ontario History* 55 (March 1963): 15 – 21.

Marsh, J. S. "Flora and Fauna." In *The Geography of the Peterborough Area*. Peterborough, ON: Trent University, Department of Geography, 1972.

Matthews, John. "Literature and Environment: Inheritance and Adaptation—The Canadian Experience." In *Commonwealth Literature: Unity and Diversity in a Common Culture*, edited by John Press. London: Heineman Education Books, 1964.

McIllwraith, Thomas F. "Transportation in the Landscape of Early Upper Canada." In *Perspectives on Landscape and Settlement in Nineteenth Century Ontario*, edited by J. David Wood. Ottawa: Carleton University Press, 1975.

McNally, David. "Staples Theory as Commodity Fetishism: Marx, Innis, and Canadian Political Economy." *Studies in Political Economy* 6 (August 1981): 35 – 63.

Miller, Harry. "Potash from Wood Ashes: Frontier Technology in Canada and the United States." *Technology and Culture* 21 (April 1980): 187 – 208.

Morton, W. L. "Clio in Canada: The Interpretation of Canadian History." *University of Toronto Quarterly* 15 (April 1946): 227 – 34.

———. "Victorian Canada." In *The Shield of Achilles: Aspects of Canada in the Victorian Age*, edited by W. L. Morton. Toronto: McClelland and Stewart, 1968.

Murray, Florence B. "Agricultural Settlement on the Canadian Shield: Ottawa River to Georgian Bay." In *Profiles of a Province: Studies in the History of Ontario*, edited by Edith Firth. Toronto: Ontario Historical Society, 1967.

Palmer, Bryan D. "Discordant Music: Charivaris and Whitecapping in Nineteenth-Century North America." *Labour/Le Travailleur* 3 (1978): 5 – 62.

Parker, Keith A. "Colonization Roads and Commercial Policy." *OH* 67 (March 1975): 31 – 8.

Parson, Helen E. "The Colonization of the Southern Canadian Shield in Ontario: The Hastings Road." *OH* 79 (September 1987): 265 – 73.

Peterman, Michael A. "'Splendid Anachronism': The Record of Catharine Parr Traill's Struggles as an Amateur Botanist in Nineteenth-Century Canada." In *Re(Dis)covering Our Foremothers: Nineteenth-Century Canadian Women Writers*, edited by Lorraine McMullen. Ottawa: University of Ottawa, 1990.

Pratt, Larry, and Matina Karvellas. "Nature and Nation: Herder, Myth and Cultural Nationalism in English Canada." *National History* 1 (Winter 1997): 59 – 77.

Richards, J. Howard. "Lands and Policies: Attitudes and Controls in the Alienation of Lands in Ontario During the First Century of Settlement." *OH* 50 (Autumn 1958): 193 – 209.

Roberts, William I., III. "American Potash Manufacture Before the American Revolution." American Philosophic Society, *Proceedings* 116 (October 1972): 383 – 95.

Rogers, Edward S. "The Algonquian Farmers of Southern Ontario, 1830 – 1945." In *Aboriginal Ontario: Historical Perspectives on the First Nations*, edited by Edward S. Rogers and Donald B. Smith. Toronto: Dundurn Press, for the Ontario Historical Studies Series, 1994.

Romney, Paul. "From the Types Riot to the Rebellion: Elite Ideology, Anti-Legal Sentiment, Political Violence, and the Rule of Law in Upper Canada." *OH* 79 (June 1987): 113 – 44.

Russell, Peter A. "Forest into Farmland: Upper Canadian Clearing Rates, 1822 – 1839." *Agricultural History* 57 (July 1983): 326 – 39.

Ryan, Pauline. "A Study of Irish Immigration to North Hastings County." *OH* 83 (March 1991): 23 – 37.

Saunders, R. M. "The First Introduction of European Plants and Animals into Canada." *CHR* 16 (December 1935): 388 – 406.

See, Scott W. "Nineteenth-Century Collective Violence: Toward a North American Context." *Labour/Le Travail* 39 (Spring 1997): 13 – 38.

Smyth, William J. "Social, Economic and Landscape Transformations in County Cork from the Mid-Eighteenth to the Mid-Nineteenth Century." In *Cork History and Society: Interdisciplinary Essays on the History of an Irish County*, edited by Patrick O'Flanagan and Cornelius G. Buttimer. Dublin: Geography Publications, 1993.

Spragge, George W. "Colonization Roads in Canada West, 1850 – 1867." *Ontario History Papers and Records* 49 (Winter 1957): 1 – 17.

Stanzel, Franz K. "Innocent Eyes?: Canadian Landscape as Seen by Frances Brooke, Susanna Moodie, and Others." *International Journal of Canadian Studies/Revue internationale d'études canadiennes* 4 (Fall/Automne 1991): 97 – 109.

Surtees, R. J. "The Development of an Indian Reserve Policy in Canada." In *Historical Essays on Upper Canada*, edited by J. K. Johnson. Toronto: McClelland and Stewart, Ltd., 1975.

Taylor, Alan. "The Great Change Begins: Settling the Forest of Central New York." *New York History* 76 (July 1995): 265 – 90.

Taylor, Colin. "The Hydrology of Peterborough and the Kawarthas." In *Peterborough and the Kawarthas*, edited by Peter Adams and Colin Taylor. Peterborough, ON: Heritage Publications, 1985.

Theberge, Elaine. "The Untrodden Earth: Early Nature Writing in Canada." *Nature Canada* 3 (1974): 30 – 6.
Upton, L.F.S. "The Origins of Canadian Indian Policy." *Journal of Canadian Studies* 8 (November 1973): 51 – 61.
Wallis, Hugh M. "James Wallis, Founder of Fenelon Falls and Pioneer in the Early Development of Peterborough." *OH* 53 (December 1961): 257 – 71.
Watson, Harry L. "'The Common Rights of Mankind': Subsistence, Shad, and Commerce in the Early Republican South." *Journal of American History* 83 (June 1996): 13 – 43.
Weaver, John. "Crime, Public Order, and Repression; The Gore District in Upheaval, 1832 – 1851." *OH* 78 (September 1986): 175 – 207.
Wurtele, D. J. "Mossom Boyd: Lumber King of the Trent Valley." *OH* 50 (Autumn 1958): 177 – 89.
Wylie, William N. T. "Poverty, Distress, and Disease: Labour and the Construction of the Rideau Canal, 1826 – 32." *Labour/Le Travilleur* 11 (Spring 1983): 7 – 29.
Wynn, Graeme. "Notes on Society and Environment in Old Ontario." *Journal of Social History* 13 (Fall 1979): 49 – 65.
———. "On the Margins of Empire (1760 – 1840)." In *The Illustrated History of Canada*, edited by Craig Brown. Toronto: Key Porter Books, 1997.
Zeller, Suzanne. "The Spirit of Bacon: Science and Self-Perception in the Hudson's Bay Company, 1830 – 1870." *Scientia Canadensis* 13 (Fall/Winter 1989): 79 – 101.

Unpublished Theses and Papers

Anderson, Richard B. "Making Wilderness Smile: Professional Resource Evaluation in Victorian Canada." Ph.D. diss., York University, 1992.
Cook, Ramsay. "Canada: An Environment Without a History?" Paper presented at "Themes and Issues in North American Environmental History." University of Toronto, 24 – 26 April 1998.
Macdonald, John Andrew Leonard. "A Region's Roots: Settlement and Land Alienation Along the Bobcaygeon Road, 1858 – 1900." Master's thesis, York University, 1994.
Stuart, Richard. "'History Is Concerned with More Than Ecology But It Is Concerned with Ecology Too': Environmental History in Parks Canada." Canadian Historical Association Meeting, 1994. Mimeographed.
Wilson, Barbara E. "'Strangers in a Strange Land'—Literary Use of Canadian Landscape by Five Genteel Settlers." Master's thesis, University of Guelph, 1973.

Index

G

H

I

J

K

L

M

N

T

U

V